SpringerBriefs in Computer Science

SpringerBriefs present concise summaries of cutting-edge research and practical applications across a wide spectrum of fields. Featuring compact volumes of 50 to 125 pages, the series covers a range of content from professional to academic.

Typical topics might include:

- A timely report of state-of-the art analytical techniques
- A bridge between new research results, as published in journal articles, and a contextual literature review
- A snapshot of a hot or emerging topic
- An in-depth case study or clinical example
- A presentation of core concepts that students must understand in order to make independent contributions.

Briefs allow authors to present their ideas and readers to absorb them with minimal time investment. Briefs will be published as part of Springer's eBook collection, with millions of users worldwide. In addition, Briefs will be available for individual print and electronic purchase. Briefs are characterized by fast, global electronic dissemination, standard publishing contracts, easy-to-use manuscript preparation and formatting guidelines, and expedited production schedules. We aim for publication 8–12 weeks after acceptance. Both solicited and unsolicited manuscripts are considered for publication in this series.

**Indexing: This series is indexed in Scopus, Ei-Compendex, and zbMATH **

Cristian Axenie · Meisam Akbarzadeh ·
Michail A. Makridis · Matteo Saveriano ·
Alexandru Stancu

Applied Antifragility in Technical Systems

From Principles to Applications

With a Foreword by Nassim Taleb

 Springer

Cristian Axenie
Department of Computer Science
and Center for Artificial Intelligence
Technische Hochschule Nürnberg Georg
Simon Ohm
Nuremberg, Bayern, Germany

Michail A. Makridis
Traffic Engineering and Control
ETH Zurich
Zürich, Switzerland

Alexandru Stancu
Department of Electrical and Electronic
Engineering
The University of Manchester
Manchester, UK

Meisam Akbarzadeh
Department of Transportation Engineering
Isfahan University of Technology
Isfahan, Iran

Matteo Saveriano
Department of Industrial Engineering
University of Trento
Povo, Trento, Italy

ISSN 2191-5768   ISSN 2191-5776 (electronic)
SpringerBriefs in Computer Science
ISBN 978-3-031-90424-0   ISBN 978-3-031-90425-7 (eBook)
https://doi.org/10.1007/978-3-031-90425-7

© The Editor(s) (if applicable) and The Author(s), under exclusive license to Springer Nature Switzerland AG 2026

This work is subject to copyright. All rights are solely and exclusively licensed by the Publisher, whether the whole or part of the material is concerned, specifically the rights of translation, reprinting, reuse of illustrations, recitation, broadcasting, reproduction on microfilms or in any other physical way, and transmission or information storage and retrieval, electronic adaptation, computer software, or by similar or dissimilar methodology now known or hereafter developed.
The use of general descriptive names, registered names, trademarks, service marks, etc. in this publication does not imply, even in the absence of a specific statement, that such names are exempt from the relevant protective laws and regulations and therefore free for general use.
The publisher, the authors and the editors are safe to assume that the advice and information in this book are believed to be true and accurate at the date of publication. Neither the publisher nor the authors or the editors give a warranty, expressed or implied, with respect to the material contained herein or for any errors or omissions that may have been made. The publisher remains neutral with regard to jurisdictional claims in published maps and institutional affiliations.

This Springer imprint is published by the registered company Springer Nature Switzerland AG
The registered company address is: Gewerbestrasse 11, 6330 Cham, Switzerland

If disposing of this product, please recycle the paper.

This book is about designing and building antifragile technical systems that turn uncertainty, randomness, and volatility to lasting advantage.

Foreword

What is Antifragility?

Abstract
Antifragility is a unifying mathematical modeling framework transferring properties from the functional domain of dose-response into the probabilistic one in distribution of outcomes, and vice versa.

Consider the following seemingly disconnected or loosely connected classes of natural and human phenomena:

Class 1 *Upregulation: Effects related to benefiting from stressors.* These include hormesis and hypertrophy in medicine, post-traumatic growth in psychology, tumor resistance in oncology, hydra-like outcomes in mythology, as well as popular beliefs about rebounds from adversity.

Class 2 *Philostochasticity: Effects related to benefiting from variance and dispersion.* These include stochastic resonance in physics and signal processing, intermittent fasting and variable dosing in medicine, "long" volatility in finance. Evolutionary processes require a certain dose of noise, variance, or replication error to satisfy a diversity of outcomes, with the hope that some of the resulting offspring will be more adapted to the environment.

Class 3 *Scaling: Effects related to allometry.* These include optimal size of animals, cities, and corporations, the fragility induced by an increase in size (stochastic diseconomies of scale), the behavior of biological entities at different scales.

Now note the property of items with opposite qualitative attributes.

Class 4 *Fragility: Effects related to breaking or rupturing under shocks and stressors at some intensity.*

Class 5 *Short volatility: harm by dispersion of outcomes and second-order effects at some window or time interval.* For example eating continuously might be harmful, but intermittently can only benefit at some time window (while a daily or alternate day occurrence may help, a monthly one can be deadly).

Class 6 *Effects linked to hazards associated with the passage of time*: it includes decay from memoryless shocks, aging, ruin probabilities, and absorbing barriers.

The idea behind antifragility isn't a descriptive approach to these attributes, nor an uncovering of these well observed phenomena, but a unifying mathematical modeling framework integrating all these classes, and, centrally **transferring properties from the functional domain of dose-response into the probabilistic one**.

Let y be the response, $y = f(x)$ a function of X a random (or deterministic) variable; we are concerned with $f(x)$; and the nonlinearity of f is determinant in altering the statistical properties of x. Functions in most applications are piecewise convex or concave, giving a rich set of responses—in general, the more nonlinear f, the more its outcomes will be divorced from the statistical properties of X.

Critically, we may not observe the full properties of X, owing to statistical incompleteness, idiosyncratic behavior, and sample insufficiency; but we can certainly assess the behavior of f via perturbation methods, in the body and the tails of the distribution. We can even sometimes influence the function, a method dubbed "convexification" or "tail clipping" applicable in finance. The idea of contracts is to eliminate, share, or transfer parts of the distribution, which alters the probability distribution of f.

The entire concept is based on a definition of fragility, which in [1, 35] is grounded in the following property. Fragility, for probabilistic reasons, must be accompanied with an accelerated response to harm, as the cumulative effect of regular, high-frequency events must be smaller in effect than those stemming from the tails of the distribution. This is a selection effect akin to the survivorship bias: being linear to harm would necessarily break the object under an ordinary intensity of stressors; what has survived must be nonlinear, having a milder response in the body of the distribution and a stronger one away from the center. Estimating the effects of tail risk resides in the nonlinearity of the response with respect to tail events [2], thus facilitating robust stress testing by focusing on acceleration rather than just magnitude.

Further, an accelerating (super-linear) response to negative stressors (as well as the passage of time) and a decelerating (sub-linear) response to positive outcomes portend fragility; the reverse situations represent antifragility, limited of course to a specific range of variations and a certain time window.

One can be fragile outside a range of variation, anti-fragile inside (though not the opposite). For antifragility is not the mirror opposite of fragility. Irreparable breaking is an absorbing barrier, which stops the unit at the point of non-recoverable ruin. The antifragile does not get absorbed in a similar manner; the asymmetry generates analytical difficulties. We also note that fragility and antifragility are associated with a *specific* source of variation. Natural systems, particularly biological ones, are universally nonlinear in their responses (sometimes extreme where the effect changes in sign, as reflected by the expression "the dose makes the poison"); hence, they lend themselves to analyses translating from the functional to the probabilistic and reciprocally, inviting a spate of medical applications [1].

One area of research with great potential can exploit the property that the transfer from the functional to the probabilistic can also take place in the reverse direction. One such prospective medical application is figuring out the frequency of past famines or shortages in food groups (say, protein) in the habitat from which a certain human groups was adapted. To take a simple application: it can consist in assessing how the intermittence of feeding increases or decreases insulin sensitivity or some other target metric, or assessing the optimal frequency of deprivation for autophagy. The method can also shed some light on the process of aging from the mismatch between lifestyle and ancestral statistical properties, in addition to the root of many diseases stemming from the deprivation of stressors.

We note that both what we call "technical systems", that is, largely manmade and engineered, and the "natural" ones, that is, largely biological, share the same properties; this likely stems from their partaking of the same attributes of nonlinearities, particularly when looked upon dynamically rather than by using comparative statics.

This volume explores the rich sets of outcomes that result from the investigation of the properties of the function $f(.)$ in a variety of domains.

Atlanta, GA, USA   Nassim Nicholas Taleb

References

1. Taleb, N. N., & West, J. (2023). Working with convex responses: Antifragility from finance to oncology. *Entropy, 25*(2), 343.
2. N. N. Taleb., Canetti, E., Kinda, T., Loukoianova, E. & Schmieder, C. A new heuristic measure of fragility and tail risks: application to stress testing. *International Monetary Fund*, 2013.

Preface

This book's idea and goal are nothing more than a scientific tour-de-force that started just 10 years after Nassim Taleb introduced antifragility. Pushing the boundaries of the conceptual work Taleb carried in his book, we go interdisciplinary. Using dynamical systems as a common language, we analyzed, developed, and collected insights into how engineered technical (dynamical) systems can turn antifragile. As Taleb described in the foreword to this work, a central goal of our initiative is to provide a mathematical framework to translate intrinsic properties within the function domain into the probabilistic domain. We hereby challenge the engineering community to rethink its design patterns towards considering intrinsic, inherited, and induced antifragility as analysis tools and requirements in any new design. We build upon Taleb's framework to provide cross-domain "recipes" to detect and quantify antifragility and "guidelines" to build antifragile systems. In a rather "engineered" approach, we propose a modular framework for both the analysis and synthesis of antifragile systems, covering robotics, traffic control, and process engineering. We hope this initial effort will motivate engineering systems communities to consider antifragility as a "first-class citizen" across the spectrum of behavior their canonical systems exhibit.

Nuremberg, Germany	Cristian Axenie
Isfahan, Iran	Meisam Akbarzadeh
Zürich, Switzerland	Michail A. Makridis
Trento, Italy	Matteo Saveriano
Manchester, UK	Alexandru Stancu
February 2025	

Acknowledgements The Applied Antifragility Research Group would like to thank Mr. Nassim Nicholas Taleb for formalizing the concept of antifragility and for the multiple interactions we had that helped us to refine the applied dimension of this framework. The group's mission is to build a foundational knowledge base by applying antifragile system design, analysis, and development across domains. We are interested in formalizing principles and an apparatus that turns the basic concept of antifragility into a tool for designing and building closed-loop systems that behave beyond robust in the face of uncertainty, volatility, and stress. We are a very diverse group of researchers from different fields. Such an interdisciplinary constellation allows us to explore applied antifragility through multiple lenses and take various research paths. Our latest research is curated and featured on the group's website at https://www.antifragility.science/.

Meisam Akbarzadeh was supported by the ETH Zürich Risk Center and partly by the Alexander von Humboldt Foundation (through a Georg Forster Fellowship) for his research while writing this book.

Competing Interests The authors have no competing interests to declare that are relevant to the content of this manuscript.

Contents

1 **Introduction** 1
  1.1 Definitions 1
  1.2 Intrinsic Antifragility 2
  1.3 Inherited Antifragility 2
  1.4 Induced Antifragility 3
  References 3

2 **Multistability and Intrinsic Antifragility** 5
  2.1 Multistability 5
  2.2 Definitions of Stability 6
    2.2.1 Linear Stability 6
    2.2.2 Basin Stability 7
  2.3 Mechanisms Leading to Antifragile Multistability 7
  2.4 Detecting Antifragile Multistability 8
  References 8

3 **Inherited Antifragility** 11
  3.1 Inherited Antifragility in Traffic Systems 11
  3.2 Discussing the Fragility of Road Transport Systems 14
    3.2.1 Traffic State Understanding 16
    3.2.2 Fundamental Diagrams and Disruptions 17
    3.2.3 Model-Based Mathematical Fragility Analysis 19
  3.3 Inherited Antifragility in Traffic Management 27
    3.3.1 Reinforcement Learning in Traffic Operations 27
    3.3.2 Antifragile Perimeter Control 28
    3.3.3 Reinforcement Learning Mechanism 32
    3.3.4 Antifragile Reinforcement Learning 34
    3.3.5 Results Showing Inherited Antifragility 37
    3.3.6 Conclusions 39
  References 40

4 Induced Antifragility ... 45
4.1 Antifragile Road Traffic Control ... 45
4.1.1 Introduction ... 45
4.1.2 Induced Traffic Antifragility ... 50
4.1.3 Antifragile Traffic Control ... 52
4.1.4 Experimental Results ... 63
4.1.5 Discussion ... 68
4.1.6 Conclusion ... 68
4.2 Antifragile Robotic Systems ... 69
4.2.1 Trajectory Tracking for Wheeled Mobile Robots ... 69
4.2.2 Wheeled Robot Modelling and Control ... 72
4.2.3 Antifragile Control of Nonholonomic Robots ... 77
4.2.4 Conclusion ... 90
4.3 Antifragile Stability Analysis of Nonlinear Control Systems ... 90
4.3.1 Preliminaries ... 91
4.4 Antifragile Stability ... 92
4.4.1 Antifragile Stability for Time-Invariant Dynamic Systems ... 92
4.4.2 Consequence of the Lyapunov Theorem ... 93
4.4.3 AF-Stability for Time-Invariant Differential Inclusions ... 95
4.5 Stability Analysis for Time-Dependent Systems and Time-Dependent Differential Inclusions ... 97
4.5.1 Antifragile Tubes ... 97
4.5.2 AF-Stability for Time-Dependent Systems S_f and S_F ... 101
4.5.3 Safe Antifragile Tubes ... 105
4.6 Applications Examples ... 106
4.6.1 Proving Stability for Trajectory-Tracking for a Sailboat Robot ... 106
4.7 Conclusions ... 112
References ... 113

5 Conclusions and Open Research Questions ... 119
5.1 Intrinsic (Anti)-Fragility ... 119
5.2 Inherited (Anti)-Fragility ... 120
5.3 Induced (Anti)-Fragility ... 121
References ... 121

Chapter 1
Introduction

Abstract The term 'antifragility' is defined precisely as the capacity of a system to respond to input variability in a beneficial manner. This chapter briefly introduces the properties of antifragility in technical systems. We discuss the terminology and the apparatus to analyze, model, and control antifragile technical systems.

1.1 Definitions

The concept of antifragility [1] pertains to the benefit derived from the variability inherent to a dynamical system, particularly in response to environmental perturbations. Systems may respond poorly to perturbations, exhibiting a fragile character or benefit from perturbations, displaying antifragile characteristics. As illustrated in Fig. 1.1, the behavior spectrum unfolds upon multiple levels and response types, each with a distinct capacity to absorb changes, volatile onset disturbances, or unknown amplitude disturbances.

This book presents a series of methodologies for the identification, analysis and integration of antifragile behaviors within technical systems. To ensure the book's appeal across various disciplines, we have provided insights and a review of the various applications of antifragility theory in technical systems, including traffic control, robotics and process engineering. Although there is considerable overlap in the methods used to quantify and apply antifragility across disciplines, there is a need for precise definitions of the scales at which antifragility operates [2]. Therefore, we provide a concise overview of the characteristics of antifragility in applied systems and a review of the pertinent literature on the antifragility of technical systems. The book is structured around three scales that are common to technical systems: intrinsic (mechanistic input-output non-linearity analysis), inherited (evolution under extrinsic environmental signals), and induced (closed-loop feedback control). The objective is to guide the reader through the spectrum of fragility-adaptiveness-resilience-robustness-antifragility, to explain the principles behind it, and to illustrate its practical implications.

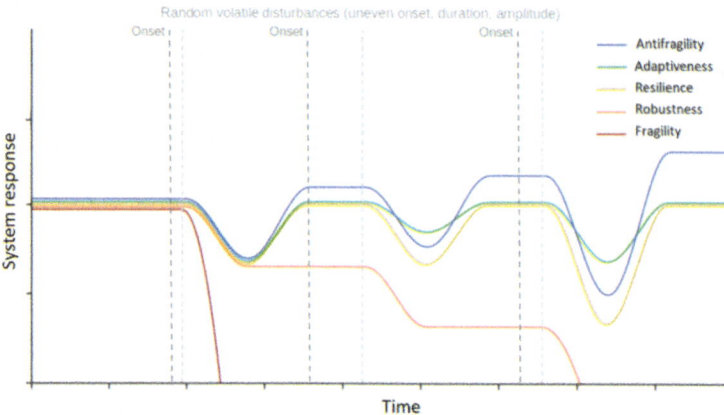

Fig. 1.1 A qualitative view on antifragile behavior and the other members of the behaviors spectrum of a dynamical system

1.2 Intrinsic Antifragility

The initial scale is that of intrinsic antifragility. In this context, the superposition of all the properties, states, and internal interactions of a system describes its response to uncertain and volatile adverse events. It is typical of technical systems to be heterogeneous, comprising a sum of well-defined payoff functions that describe the response dynamics of their composing elements. In other words, intrinsic antifragility describes the relationship between a system and volatile environmental perturbations to which the system is exposed without considering any feedback. Systems that are intrinsically antifragile benefit from an uneven distribution of internal dynamics. This is based on the convexity of the response function of the system in question, which is without external input and solely based on the internal components' heterogeneity and resilience. Features such as stability describe the most simple system response with minimal antifragile characteristics. Within this scale, precise characterization of the payoff function describing the relationship between system inputs and outputs is of the utmost importance.

1.3 Inherited Antifragility

The subsequent scale of antifragility is that of inherited antifragility. Inherited antifragility is defined by the additional interactions that a system has with other connected systems, as indicated by the term itself. The degree of homogeneity and/or heterogeneity within a system is of significant importance concerning the design and synthesis of inherited antifragile behaviors. From criticality and multi-level interactions of multiple timescales to quantifying criticality margins, the design scheme in question leverages local interactions of the system to facilitate benefit from perturbations. In other words, inherited antifragile systems benefit from input distribution unevenness, based on the emergent system dynamics and its interactions with the operating environment (i.e. disturbances, noise, modulated perturbations). Finally, the pattern of interaction also dictates the harmonization among the intrinsic behavior of the system and the externally coupled, uncontrollable modulation of its peers. Interestingly, it is this interaction among scales which can push the system towards antifragility.

1.4 Induced Antifragility

The highest level of antifragility assumes not only external influence on the system dynamics (as in the inherited case) but also external goal-oriented guidance. In this case, an entity that embeds knowledge of the system's states, an approximation of its external stressors, and the goal of the system dynamics, generates signals that drive the system towards a prescribed location in its state space in a judicious manner. The induction of a desired behavior within induced antifragility necessitates the implementation of a feedback control-theoretic, design and synthesis approach. In this approach, nonlinear dynamics across both space and time can enhance the system's capacity to absorb internal and external disruptions. Induced antifragile systems benefit from input distribution unevenness based on emergent system dynamics in closed-loop with a controller driving the system towards prescribed dynamics in the presence of modulated or non-stationary disturbances, noise, and volatility.

References

1. Taleb, N. N. (2012). *Antifragile: Things that gain from disorder* volume 3. Random House.
2. Axenie, C., Lopez-Corona, O., Makridis, M., Akbarzadeh, M., Saveriano, M., Stancu, A. & West, J. Antifragility as a complex system's response to perturbations, volatility, and time. *ArXiv Preprint* ArXiv:2312.13991. (2023)

Chapter 2
Multistability and Intrinsic Antifragility

Abstract This chapter introduces intrinsic antifragility and its connection with the multistability of a technical dynamical system. From the mathematical apparatus of criticality and multistability, we delve into antifragility stability. We consider methods for the detection, analysis, and modelling of complex large-scale systems antifragility.

2.1 Multistability

Antifragility is all about moving to a "better" equilibrium due to external stress. Hence, the existence of more than one steady state must be established before one can discuss the intrinsic antifragility of a system. Various kinds of steady states, including rare attractors, hidden attractors, self-excited attractors, and strange attractors, would satisfy the requirement of multiple equilibria. Nevertheless, the question about the antifragility of metastable systems should be addressed in further investigations.

The multistability of a system is a prerequisite for its intrinsic antifragility. However, establishing multistability within systems can be complex. This chapter explores the different types of multistability and various definitions of stability. It also explains the mechanisms that lead to multistability, providing system analysts with insights into what to examine in their respective systems. While multistability may be undesirable in many technical systems, it remains essential for achieving intrinsic antifragility.

Multistability is manifested through bifurcation in the phase space of the system. Bifurcation points are the parameter values of the system at which new steady states are produced, vanish, or the type of stability of a steady state changes. Bifurcations where new steady states are produced may be either local or global. When the control parameter crosses a critical point and hence changes the local stability—defined in the next subsection—of a fixed point or a periodic orbit, a local bifurcation occurs. On the other hand, when an invariant set collides with another set or with an equilibrium point, global bifurcation appears. In global bifurcation, the qualitative structure of the phase space changes.

Various types of bifurcation include pitchfork bifurcation (consisting of super- and subcritical bifurcations), saddle-node bifurcation, transcritical bifurcation, Hopf bifurcation (consisting of super- and subcritical Hopf bifurcations), Neimark–Sacker bifurcation, multiple limit cycle bifurcation, infinite-period bifurcation, inverse gluing bifurcation, and symmetry-increasing bifurcation.

In the following subsections, we explain various types of stability and then discuss the causes and manifestations of the multistability of systems. We first describe various types of stability and then discuss the origins of multistability followed by the manifestation of multistability in technical systems.

2.2 Definitions of Stability

Before diving into the concept of stability, we need to clarify the concept of equilibrium. There are various terms used to describe a stable system, i.e., its state is not changing over time. Fixed point, equilibrium, and steady-state are the terms used for explaining the situation in which a system is not changing over time. A fixed point is a value that remains unchanged under a given function. In other words, if f is a function, a point x is a fixed point if $f(x) = x$. In the context of dynamical systems, equilibrium refers to a state where the system has no net change over time. For a system of differential equations, an equilibrium is a state where all derivatives are zero. Mathematically, if the system is described by $\frac{dx}{dt} = f(x)$, then an equilibrium point x_e satisfies $f(x_e) = 0$. Steady-state is a concept often used for physics and engineering systems that are subject to external inputs or boundary conditions. A system is said to be in a steady state if its variables do not change over time, despite the probable existence of some ongoing processes within the system.

2.2.1 Linear Stability

An equilibrium (and generally an invariant set) is stable if the trajectories initiated in its proximity stay close to it. Various formulations for stability have been proposed since the concept was initiated. A system is Lyapunov stable if its phase-space trajectory fluctuates close to an equilibrium point. In other words, a system is Lyapunov stable concerning x^* if for every $\epsilon > 0$ and $t_0 \geq 0$ there exists $\delta > 0$ such that from $||x(t_0) - x^*|| < \epsilon$ one can conclude $||x(t) - x^*|| < \delta$ for all t after t_0.

If the phase space trajectory of a system gradually approaches an equilibrium, the system is called asymptotically stable. Mathematically, if there exists $\delta \geq 0$ such that $|x(t) - x^*| \leq \delta$, it implies that $|x(t) - x^*|$ merges to zero as time passes, and the equilibrium is asymptotically stable. The asymptotic stability may be local or global. An equilibrium of the system is locally asymptotically stable if it is Lyapunov stable and there exists $M \geq 0$ such that $|x(t_0) - x^*| \leq M$ implies that $\lim_{t \to \infty} x(t) = x^*$. The equilibrium is globally asymptotically stable if the criterion holds for $M \geq 0$.

An equilibrium is exponentially stable if it is asymptotically stable, and all trajectories in the proximity of the equilibrium approach it exponentially. In mathematical form: $||x(t_0) - x^*|| < \delta \implies ||x(t) - x^*|| \leq ce^{-\delta t}$ for $t \gg t_0$.

For the systems having orbits in their phase space trajectory, orbital stability can be defined under the effect of small external disturbances. A periodic solution is orbitally stable if it is Lyapunov stable and for any small perturbation, the new orbits completely fall within the previous one. Structural stability considers the sensitivity of the system's structure (such as the number and type of the attractors) to small disturbances.

The integral stability is concerned with the stability of the system under constant perturbations when the perturbations are sufficiently small everywhere, probably except for a small interval [1].

2.2.2 Basin Stability

Basin stability is the fraction of the initial conditions that lead to a steady state [2]. In other words, the basin of attraction of an equilibrium can be measured as the probability that the system returns to the equilibrium after some random perturbations. The higher the probability, the higher the stability of the steady state. Basin stability complements the linear-stability paradigm [3]. Survivability is a measure of basin stability. It is the fraction of initial conditions that generate trajectories remaining in the "desired region" of the phase space for a given time [4]. The set of these initial conditions is called the basin of survival.

Besides the volume of the basin of a steady state, ecological resilience may also be used to quantify the stability. Ecological resilience is the ability of a system to get back to its current steady state after being disturbed by an external noise [5]. Four parameters are used for quantifying ecological resilience: latitude, resistance, precariousness, and panarchy. Latitude indicates how much a system can change without switching to another state. Resistance indicates the robustness of the system state to external perturbations, i.e. how large the perturbation should be to switch the system to another regime. Precariousness indicates how close the current steady state of the system is to a threshold between the different dynamical regimes. Panarchy indicates the interaction between coexisting steady states and their basins of attraction, describing how a change in one basin affects other basins [6].

2.3 Mechanisms Leading to Antifragile Multistability

Several mechanisms have been mentioned as the origins of multistability in systems. One would expect at least one of these mechanisms to exist in the system under study as a necessary (but not sufficient) requirement for intrinsic antifragility. The addition of a small amount of dissipation in a conservative system [7], formation of clusters in a system composed of two or more coupled systems [8], the coexistence of multiple synchronous states having different phase shifts between oscillations [9], existence of homo-clinic tangents [10], positive feedback [11], delayed feedback [12], periodic forcing [13], symmetry [14], and structural multistability [15] are among mechanisms that can lead to multistability.

2.4 Detecting Antifragile Multistability

To reveal the intrinsic antifragile multistability of a system, one may simply vary the initial conditions and observe the equilibrium that the system finally settles at. This way, a cell-to-cell mapping can be laid down, which relates initial conditions to steady states. This may be done experimentally or through the Monte Carlo method [16]. Applying external short pulses can reveal the antifragile multistability by pushing the phase space trajectory of the system away from its current equilibrium and giving it a chance to either return to the same equilibrium or fall into the basin of another (probably existing) attractor [17].

Applying stochastic perturbations to a system converts a multistable system to a metastable system, and therefore causes the trajectory to visit the coexisting metastable states throughout time. Visiting different metastable states and switching among them is called attractor hopping [18]. Critical velocity surfaces can be used to detect multistability in multidimensional dynamical systems [19]. For each variable $x_i(t)$, its derivative with respect to time $\dot{x}_i(t) = 0$ is called velocity. By setting $\dot{x}_i(t) = 0$, x_i—critical velocity surface is detected, and the dimension of the system reduces by one unit. By repeating this process for all variables, the intersection of critical velocity surfaces would show the various steady states of the system [20].

The continuation (homotopy) method may also be used for revealing multistability. This method is used to compute the bifurcation diagrams of multi-stable systems by producing a sequence of closely located solutions of a dynamical system [21]. In the case of rare attractors, the method of complete bifurcation group could be used to detect the multi-stability [22].

References

1. Pisarchik, A. N. & Hramov A. E. Multistability in Physical and Living Systems. Cham: Springer (2022)
2. Wiley, D.A. Strogatz, S.H. Girvan, M. The size of the sync basin. *Chaos* **16**, 015103 (2006)
3. Menck, P. Heitzig, J. Marwan, N. Kurths, J. How basin stability complements the linear-stability paradigm. *Nature Phys* **9**, pp. 89–92 (2013)
4. Hellmann, F. Schultz, P. Grabow, C. Heitzig, J. Kurths, J. Survivability of deterministic dynamical systems. *Sci Rep* **13(6)** 129654 (2016)
5. Walker, B. Holling, C.S. Carpenter, S.R. Kinzig, A. Resilience, adaptability and transformability in social-ecological systems. *Ecol Soc* **9(2)**, 5 (2004)

References

6. Allen, C.R. Angeler, D.G. Garmestani, A.S. Gunderson, L.H. Holling, C.S. Panarchy: theory and application. *Ecosystems* **17** pp. 578–589 (2014)
7. Poon, L. & Grebogi, C. Controlling complexity. *Phys Rev Lett* **75**, pp. 4023–4026 (1995)
8. Kaneko, K. Chaotic but regular posi-nega switch among coded attractors by cluster size variation. *Phys Rev Lett* **63**, pp. 219–223 (1989)
9. Dudkowski, D. & Jafari, S. & Kapitaniak, T. & Kuznetsov, N.V. & Leonov, G.A. & Prasad, A. Hidden attractors in dynamical systems. *Phys Rep* **637**, pp. 1–50 (2016)
10. Newhouse, S. Non-density of axiom A(a) on S2. *Proc AmMath Soc Sympos Pure Math* **14**, pp. 191–202 (1970)
11. Cinquin, O. & Demongeot, J. Positive and negative feedback: striking a balance between necessary antagonists. *J Theor Biol*. **216**, pp. 229–241 (2002)
12. Ikeda, K. Multiple-valued stationary state and its instability of the transmitted light by a ring cavity system. *Opt Commun* **30**, pp. 257–261 (1979)
13. Sanju Varma, V.S. Quadratic map modulated by additive periodic forcing. *Phys Rev E* **48**, pp. 1670–1675 (1993)
14. Sprott, J.C. Simplest chaotic flows with involutional symmetries. *Int J Bifurc Chaos* **24**, 1450009 (2014)
15. Bertoldi, K. Vitelli, V. Christensen, J. Van Hecke, M. Flexible mechanical metamaterials. *Nat Rev Mater* **2**, pp. 1–11 (2017)
16. Hsu, C. Cell-to-cell mapping: a method of global analysis for nonlinear systems. (Springer Science & Business Media, 2013)
17. Kaneko, K. Chaotic but regular posi-nega switch among coded attractors by cluster-size variation. *Physical Review Letters*. **63**, 219 (1989)
18. Kraut, S. & Feudel, U. Multistability, noise, and attractor hopping: The crucial role of chaotic saddles. *Physical Review E*. **66**, 015207 (2002)
19. Godara, P., Dudkowski, D., Prasad, A. & Kapitaniak, T. New topological tool for multistable dynamical systems. *Chaos: An Interdisciplinary Journal Of Nonlinear Science*. **28** (2018)
20. Akbarzadeh, M., Memarmontazerin, S., Derrible, S. & Salehi Reihani, S. The role of travel demand and network centrality on the connectivity and resilience of an urban street system. *Transportation*. **46** pp. 1127-1141 (2019)
21. Krauskopf, B. & Osinga, H. Computing invariant manifolds via the continuation of orbit segments. *Numerical Continuation Methods For Dynamical Systems: Path Following And Boundary Value Problems*. pp. 117-154 (2007)
22. Zakrzhevsky, M. 398. New concepts of nonlinear dynamics: complete bifurcation groups, protuberances, unstable periodic infinitiums and rare attractor.. *Journal Of Vibroengineering*. **10** (2008)

Chapter 3
Inherited Antifragility

Abstract This chapter introduces inherited antifragility as a gained characteristic of a technical system. It describes the benefit derived from input distribution unevenness, based on the emergent system dynamics and its interactions with the operating environment (i.e. disturbances, noise), under the premise that the intrinsic dynamics of the system are exposed to the interactions. We consider methods for the detection, analysis, and modelling of road traffic systems antifragility.

3.1 Inherited Antifragility in Traffic Systems

Fragility in the context of complex systems refers to their susceptibility to disruptions. The interdependence of their components, poor system design, the initial conditions, abrupt system state transitions, and non-linearities are only some of the possible reasons. Inherited antifragility[1] is achieved through adaptation with time. Essentially, volatility is a desired property for the system that learns from past experiences under stress and adapts to future disruptions to improve its response.

Let us imagine a city that suffers from severe congestion and at some point decides to invest in public transportation infrastructure and services. Logically, it can be expected that network users progressively shift to public transport while they experience significant delays and costs until an equilibrium is reached.

Complex systems that appear in the natural or the technical context that is discussed here are always fascinating to study and challenging to monitor and control. Intrinsic non-linearities pose challenges in modeling and understanding dynamics, making the study of the efficiency of a system during an event versus the event's intensity difficult to model and describe.

This chapter focuses on the paradigm of road transportation networks as an applied domain where inherited antifragility can significantly enhance their efficiency. It is necessary to strengthen our understanding of the dynamics that govern such complex systems. Recent advances in sensor and communication technologies allow us, for the first time, to build large real-world data volumes that offer the opportunity to

---

[1] Within this chapter, the terms *inherited antifragility* and *antifragility* are used interchangeably.

observe our road networks at scale and produce new fundamental knowledge on traffic engineering topics.

The operation of road transport systems is (phenomenally) based on a simple supply and demand principle. On the one hand, drivers want to use the existing infrastructure to move from one place to another at a given time. On the other hand, there is a system with a nominal capacity, i.e., to host a specific number of users at a given time. Nevertheless, transport systems are ruled by complex high-order dynamics. Consequently, traffic state estimation and control are challenging and intriguing. The reason behind this is the spatio-temporal heterogeneity of individual driver behaviors that interact with each other and the system (see Fig. 3.1).

The (in)efficiency of road transport systems has a huge socioeconomic impact (delays, costs, pollutants, emissions, noise). Consequently, it is essential to efficiently monitor and control their state to prevent the emergence of devastating phenomena for the users, such as traffic jams or gridlocks.

In traffic engineering, the state of a road or a network is monitored through analysis of observations (vehicle counts and speeds) or a space-time area. There are three main quantities to characterize the state of a road or a system: flow, density, and speed. The flow is measured at a given point in space in veh/h, i.e., the number of users that are served within a given period. The density is measured at a given space, e.g. a road segment, in veh/km. The speed property represents the average space-mean speed, i.e., concerning the road segment under study. These parameters are often used to represent the state of the system. As expected, the function between the efficiency over intensity is not monotonically increasing. The efficiency is bounded by the capacity of the system, i.e., the maximum number of users that the system can serve. In modeling approaches, the capacity is represented through a scalar corresponding to a specific density that is named critical density.

Self-organized criticality describes the state transition of certain complex systems when approaching a critical state without external intervention. Disruptive phe-

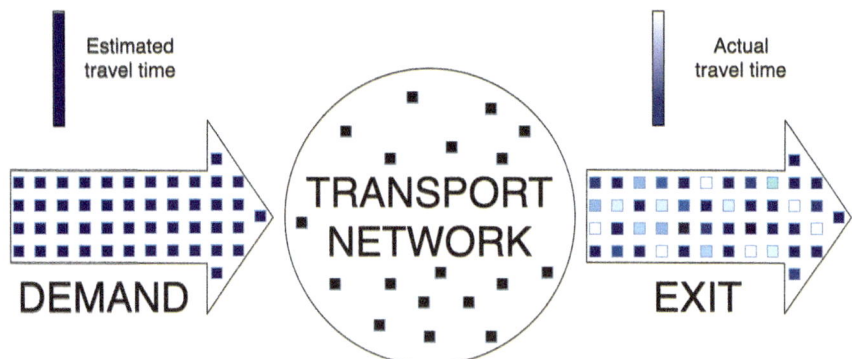

Fig. 3.1 Transport systems are based on a simple demand and supply principle. Stochasticity in the users' behavior and their interactions with other users and the infrastructure create complex traffic dynamics and deviation from expected values, i.e., in this example, the travel times

3.1 Inherited Antifragility in Traffic Systems

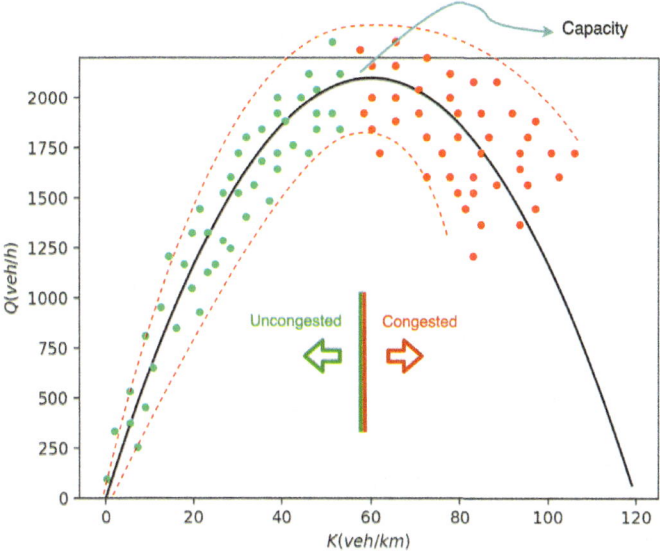

Fig. 3.2 Typical shape of a fundamental diagram describing the traffic state space of a system. The solid line represents the model, while the red dots represent typical observations. Error variance between model and empirical observations increases as the state of the network becomes more saturated, i.e. traffic congestion

nomena appear around critical states. In the transportation network paradigm, such phenomena lead to traffic jams or gridlock [1].

Although the main operation principle of transportation systems is simple (see Fig. 3.1), heterogeneity among the users of the system is the major factor for the appearance of such chaotic phenomena. In road transport, heterogeneity mainly arises during conflict resolution among network users, which in road transport is primarily due to different driver behaviors and lane-changing strategies [2, 3].

Figure 3.2 shows typical discrepancies between modeling and observations. The solid line is noted as the fundamental diagram of traffic flow. The critical density and the capacity of the system describe the optimal point in such modeling approaches. Before or after this point, the state of the system is sub-optimal. The free-flow part (on the left of the fundamental diagram) is of less interest to traffic engineering. This part describes the operation of the system when there is low density, which under normal conditions corresponds to few network users. The saturated part (on the right of the fundamental diagram) plays a major role in transport engineering. This part describes the traffic state when there is enough demand to reach the capacity of the system. Around the capacity, user conflicts lead to under-utilization of the infrastructure (in this example, the road), and hence, the service rate of the infrastructure (flow) decreases. Consequently, any strategy that aims at maximizing the efficiency of such a system pushes the state of the system near the above-mentioned optimal point.

Antifragility is a key property for in-depth system understanding and the application of any control strategy. An antifragile system learns from disruptions. In the example of Fig. 3.2 the disruption occurs when the state transits to the right part of the fundamental diagram. Antifragile design monitors by definition the state transition, i.e. the direction of the change (left or right), the speed of the change during a time interval and the acceleration of change. Direction, speed, and acceleration of state transition are often considered underrated properties in traffic engineering, yet they can contribute to the timely detection or prediction of critical disruptions. Antifragile design is based on the study and embedding of these properties in the proposed solution but also on the training of the system over time to be able to learn after every disruption and provide better responses in the future. Nevertheless, it should be noted that antifragile design creates a trade-off with stable conditions; as a system is trained on outliers, its performance logically decreases under normal operation. Nevertheless, since losses in complex systems follow a power-law distribution, the anticipated benefits under disruptions are far greater than the marginal losses under stable conditions.

3.2 Discussing the Fragility of Road Transport Systems

Road traffic grows over the years, and building new infrastructure only leads to induced traffic without resolving traffic congestion [4]. Finding intelligent ways for monitoring, state estimation, and control toward better utilization of existing supply resources is essential for maintaining a good level of service for the users. The unobstructed and efficient operation of road transport systems has become very important. Consequently, we need to understand the fragility of road networks as well as ways to fuse antifragile properties into existing complex operational systems.

Modern road transport systems, with the aid of technological advancements in sensors, software and communication, are expected to support sustainable modern cities, energy consumption minimization, encouragement of cycling and walking, and building healthier habits and more [5–9]. Empirical evidence indicates that road transportation networks exhibit fragile properties. Such an example is the well-known BPR function [10], which illustrates at the link level that travel time grows exponentially with traffic flow, leading to an infinite temporal cost when the traffic influx is at the maximal density of the network. Similar properties appear in other transportation systems, such as railways and aviation [11].

Researchers have devoted extensive efforts to the assessment and design of efficient and resilient transportation systems [12–15]. This chapter highlights the importance of systems that not only withstand disruptions but learn over time and improve their response to them [16]. This becomes increasingly topical with the availability of vast amounts of observation data [17].

Two commonly used terms to characterize the extent of performance variations under stress are robustness [18, 19] and resilience [20, 21]. However, the definitions of robustness and resilience vary across different contexts and are sometimes used

3.2 Discussing the Fragility of Road Transport Systems

interchangeably. Here, we adopt the definition proposed in [22], wherein *robustness involves evaluating a system's ability to maintain its initial state and withstand performance degradation when confronted with uncertainties and disturbances*. On the other side, resilience emphasizes a system's capability to recover from major disruptions and return to its original state. In brief, robustness relates to resistance, whereas resilience is about recovery.

Nevertheless, these two terms can overlook the consideration over a long timeline and the potential escalation of disruptions, which is particularly relevant in transportation with unforeseen disruptions on the demand side, e.g. social events, and the supply side, e.g., road works. The concept of antifragility [23, 24] aims at transforming people's understanding and perception of risk. When employed in systems and control, (anti-)fragile design focuses on the relationship between the performance and the magnitude of disruptions. A convex relationship between loss increase over increasing disturbance indicates a fragile system, while a concave one indicates an antifragile system.

An (anti-)fragile response of a system can be characterized through a nonlinear relationship between the performance loss and the magnitude of the disruption, as shown in Fig. 3.3a. Both nonlinear functions can be represented by Jensen's inequality [25], with either $E[g(X)] \geq g(E[X])$ for a fragile response or $E[g(X)] \leq g(E[X])$ for an antifragile response. This relationship can then be determined through the second derivative [26], i.e., a positive second derivative featuring a convex function and, hence, a fragile system and vice versa. It should be noted that the calculation of the derivatives is only possible when the function is continuous and differentiable, which means the underlying mathematical model representing the system needs to be known beforehand.

In transportation, there is a known convex relationship between travel time and traffic flow with empirical data, although not described with the term antifragility. The BPR function [10], as shown in Eq. (3.1) has given an intuitive example showing the fragility of the transportation systems on a link level. The BPR function has been extensively applied in the estimation of the link (route) travel time [27].

$$T = T_{ff}\left(1 + \alpha \left(\frac{q}{q_{\max}}\right)^\beta\right) \quad (3.1)$$

However, in most real-world scenarios, the mathematical function of the system is agnostic, and only discrete measurements of the system's performance are available. In this case, we can calculate the distribution skewness to determine the (anti-)fragile property of the system. A negative skewness represents the long tail pointing to the left and indicates an antifragile response, as shown in Fig. 3.3b.

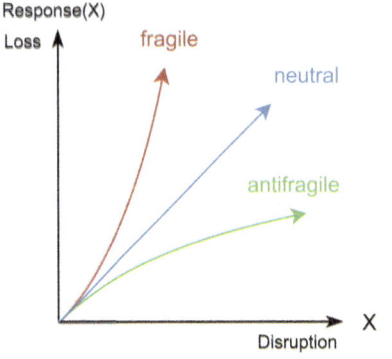 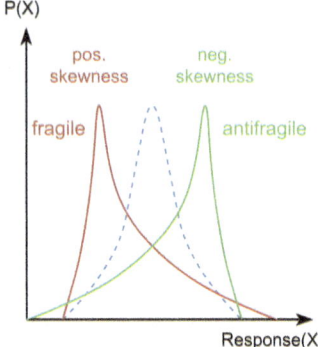

(a) Nonlinear relationship for a continuous function

(b) Distribution skewness for discrete measurements

Fig. 3.3 Characteristics and identification of (anti-)fragility

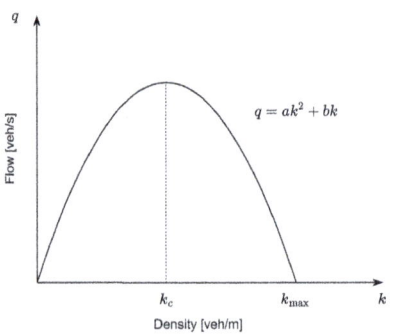 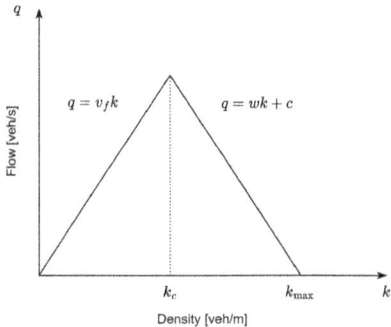

(a) Greenshields 2nd-degree polynomial FD (1934)

(b) Daganzo two-regime linear FD (1994)

Fig. 3.4 Mathematical models of FDs

3.2.1 Traffic State Understanding

The earliest study on traffic state understanding and modelling dates back to 1934 [28] on a section of a highway and yielded the first Fundamental Diagram (FD), which exhibits the relationship between traffic flow, denoted as q, and density, denoted as k, in the shape of a second-degree polynomial (see Fig. 3.4a). Later, other researchers also developed FDs in various forms, with one of the most commonly applied FDs being proposed in [29], as shown in Fig. 3.4b, which can be reproduced using only three parameters, i.e., the free-flow speed, back-propagation speed, and critical density, denoted as v_f, w, and k_c, respectively.

Similar to the FD for a road link, the Macroscopic (or Network) Fundamental Diagram (MFD) describes the space of possible traffic states within a region. Various

3.2 Discussing the Fragility of Road Transport Systems

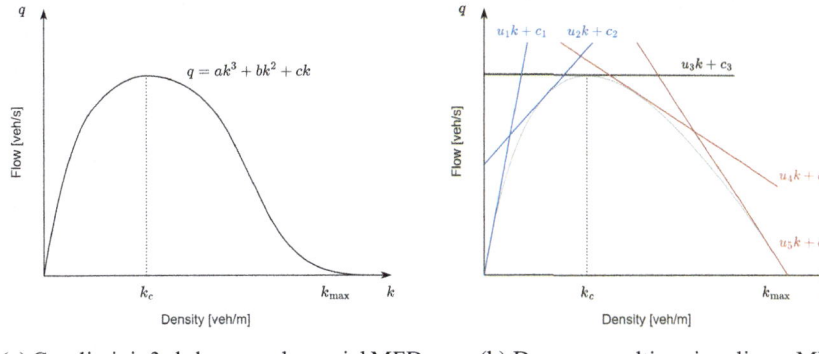

(a) Geroliminis 3rd-degree polynomial MFD (2013)

(b) Daganzo multi-regime linear MFD (2008)

Fig. 3.5 Mathematical models of MFDs

versions of MFDs exist. MFDs can be fit numerically from field measurements as a cubic polynomial, as in [30] and illustrated in Fig. 3.5a, or derived analytically exploiting variables with physical meaning (free-flow speed, lane length, etc.) as in [29] using variational theory, often referred to as the Method of Cuts (MoC). An example of multi-regime linear functions MFD from MoC is shown in Fig. 3.5b.

3.2.2 Fundamental Diagrams and Disruptions

Since traffic networks primarily involve the management of supply and demand, we consider that any traffic disruption in the real world can be classified as either a demand or a supply disruption. A demand disruption can be easily understood as, for example, surging traffic due to a social event, whereas a supply disruption may indicate an impaired network due to external factors, such as adversarial weather or lane closure.

Additionally, since disruptions represent abnormal cases that only exist temporarily, we not only consider the onset of disruptions but also assess the recovery process of the systems following such disruptions. By considering both the onset and recovery processes, it becomes possible to compare the performance between either robust and antifragile designs or resilient and antifragile designs. The scheme of onset and recovery from demand or supply disruptions are illustrated in Fig. 3.6a and b. We denote the MFD profile as $G(k)$ and assume a constant base demand in the network as q, resulting in an equilibrium traffic state in the network without any disruption.

The initial density at equilibrium, the critical density, the new density after disruption, and the gridlock density are denoted as k_0, k_c, k', and k_{max}, respectively. For the study of supply disruptions, we introduce a disruption magnitude coefficient, denoted as r, so that the disrupted MFD profile can be represented as $(1-r)G(k)$.

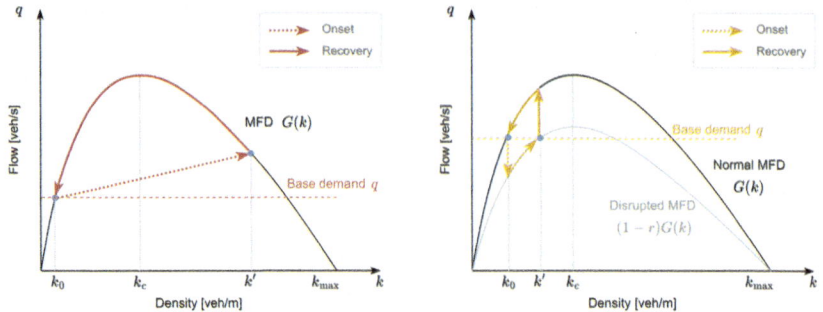

(a) Onset and recovery of a demand disruption (b) Onset and recovery of a supply disruption

Fig. 3.6 Onset and recovery of disruptions

It should also be noted that on the network level, instead of traffic flow—density, MFD can also be represented with vehicle accumulation—trip completion, such as in [31, 32]. Hence, with vehicle accumulation denoted as n, k_* and $G(k)$ can also be replace with n_* and trip completion $G(n)$.

Below, we discuss a set of assumptions that are considered for the study within this section.

Assumption 1 We study the onset of disruption,s focusing only on the two stable traffic states before and after the disruptions, while the recovery is regarded as a gradual process of congestion dissipation. □

For the onset of demand disruptions, we assume a rapid influx of traffic into the network, which can be characterized as an instantaneous event. For the onset of supply disruptions, although it may take some time for the number of vehicles within the network to accumulate, we understand it as a change from one stable traffic state $G(k_0)$ to the final stable state $(1-r)G(k')$, i.e., the equilibrium points in blue as shown in Fig. 3.6b. The process of recovery in traffic engineering typically takes longer due to the congestion dissipation process.

Assumption 2 For demand disruptions, we assume $k' > k_c$, whereas $k' < k_c$ for supply disruptions. □

A surge in demand should be considered a disruption only if it reduces the network's maximum serviceability; that is when the traffic state enters the congested zone of the MFD, and the flow drops below the network's capacity. Conversely, for a supply disruption with constant base demand, if the traffic state surpasses the maximum capacity of the disrupted MFD profile—indicating that the incoming flow exceeds the network's capacity post-disruption—the traffic density will continue to accumulate until the network reaches a full gridlock. In this scenario, no equilibrium point exists after the disruption, thereby violating Assumption 1.

3.2 Discussing the Fragility of Road Transport Systems   19

Assumption 3 For demand disruptions, we assume $q < G(k')$, whereas $q < (1-r)G(k_c)$ for supply disruptions. □

Similarly, this assumption relates to the necessity of avoiding gridlock for both demand and supply disruptions. If the base demand exceeds the outgoing flow after a demand disruption or the maximal capacity of the network after a supply disruption, then the traffic state will continue to deteriorate until complete gridlock occurs.

3.2.3 Model-Based Mathematical Fragility Analysis

This section discusses through a mathematical analysis the fragility of road transportation systems at the macroscopic level. To investigate the instantaneous disruption onset between different stable states, the Average Time Spent (ATS) indicates the performance of a network (ATS remains constant under stationary conditions). Conversely, for the examination of disruption recovery, Total Time Spent (TTS) reflect better the temporal costs for all vehicles in the process, considering that the time spent in the network varies significantly for vehicles entering at different times.

3.2.3.1 Demand Disruption

The presence of a positive second derivative in performance loss concerning the magnitude of disruption serves as an indication of the transportation system's fragility. Therefore, to illustrate the system's fragility to demand disruption, we analyze the derivatives of time spent-ATS for the onset of disruptions or TTS for the recovery process- relative to the initial disruption demand. The initial disruption demand is represented either by disruption density k' or the initial disruption demand n' when we study the relationship between trip completion and vehicle accumulation. If the system is neither fragile nor antifragile, this approach is expected to yield a linearly growing loss in performance alongside the disruption and zero derivatives. Quoting and in line with the famous statistician George Box, "All models are wrong, but some are useful", we employ both a numerical and an analytical MFD to investigate the fragile properties under the onset of demand disruption at both microscopic and macroscopic scales.

Proposition 1 *Road transportation systems are fragile with the onset of demand disruptions on the macroscopic level.* □

Proof We study the fragile property based on the third-degree polynomial MFD in [33] and multi-regime linear MFD based on MoC in [34]. For MFD approximated with a third-degree polynomial, we have:

$$G(k) = ak^3 + bk^2 + ck \tag{3.2}$$

$$v(k) = \frac{q}{k} = ak^2 + bk + c \tag{3.3}$$

Fig. 3.7 Daganzo multi-regime linear MFD

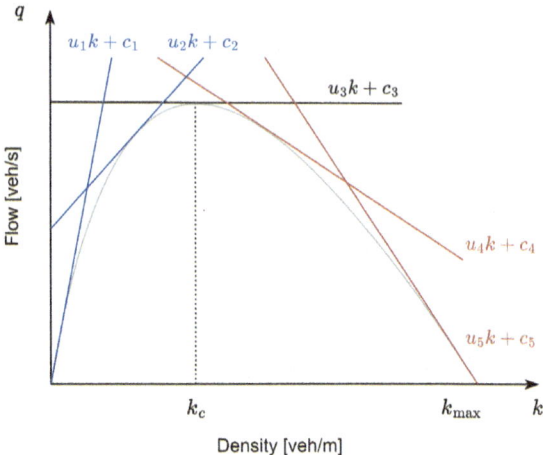

Consequently, the ATS and its first and second derivatives are:

$$ATS = \frac{L}{v(k')} = \frac{L}{ak'^2 + bk' + c} \tag{3.4}$$

$$\frac{dATS}{dk'} = \frac{-(2ak' + b)L}{(ak'^2 + bk' + c)^2} \tag{3.5}$$

$$\frac{d^2ATS}{dk'^2} = \frac{\frac{3}{2}(2ak' + b)^2 + \frac{1}{2}(b^2 - 4ac)}{(ak'^2 + bk' + c)^3} L \tag{3.6}$$

The average speed has to be a real number, indicating that Eq. (3.3) should have real roots, so $b^2 - 4ac > 0$ should hold. Therefore, the derivatives are positive.

The MFD derived with MoC can be approximated by a series of linear functions. Likewise to the Daganzo two-regime linear FD as in Eq. (3.7), for any linear function, the second derivative is:

$$\frac{d^2ATS}{dk'^2} = \frac{-2u_i c_i L}{(u_i k' + c_i)^3} \tag{3.7}$$

As coefficient c_i is positive for any cut since the y-intercept should always be positive by the definition of MoC, whether the second derivative is positive or negative depends solely on u_i. The cuts that intercept the MFD before the critical density k_c exhibit antifragile properties ($u_i > 0$, in blue as shown in Fig. 3.7) while the others with intercepts larger than the critical density k_c show fragile responses ($u_i < 0$, in red), with an exception in case there's a cut at the critical density ($u_i = 0$, in grey). Conforming to Assumption 2, for demand disruptions, we focus on the cuts with intercepts larger than the critical density k_c ($u_i < 0$). The second derivative for these cuts is positive, indicating a fragile property. □

3.2 Discussing the Fragility of Road Transport Systems

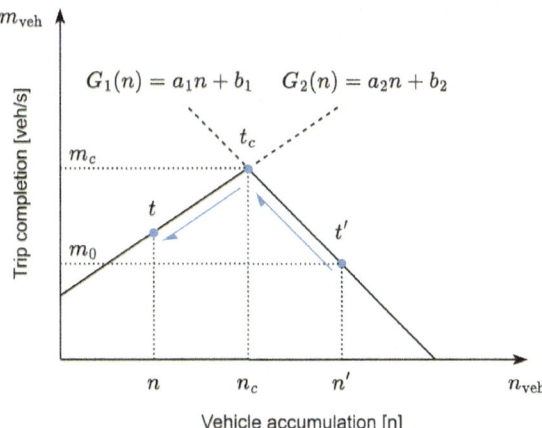

Fig. 3.8 Simplification of MoC

Proposition 2 *Road transportation systems are fragile when going through the recovery process from demand disruptions.* □

Proof According to Assumption 1, we simplify this surging demand as a disruption that takes place instantly in the network, denoted as n' at time $t' = 0$. As the MoC is composed of a series of linear functions with decreasing gradients as the vehicle accumulation increases, the multi-linear regimes can be represented into multiple sets of consecutive duo linear functions as Fig. 3.8 shows. The four constants a_1, a_2, b_1, and b_2 are the slope and y-intercept on the coordinates for the two cuts, with $a_2 > a_1$ and $b_1 > b_2 > 0$. We refer to these two linear functions as the more congested branch and the less congested branch. Also, the critical accumulation n_c here does not represent the critical point of the entire MFD, but rather the critical accumulation of any two consecutive cuts. After a certain period t_c, the number of vehicles in the network reaches this critical accumulation n_c. And after any period $t > t_c$, the vehicle accumulation becomes n. We also denote the initial trip completion and critical trip completion as $m_0 = a_1 n' + b_1$ and $m_c = a_1 n_c + b_1 = a_2 n_c + b_2$, respectively.

Any two consecutive cuts of the MFD can be formulated into the following Eq. (3.8):

$$G(n) = \begin{cases} a_1 n + b_1, & n_c \leq n < n_{\max}, \\ a_2 n + b_2, & 0 \leq n < n_c. \end{cases} \quad (3.8)$$

The system dynamics can be summarized as Eq. (3.9).

$$\frac{dn}{dt} = -G(n) + q = -a_i n - b_i + q \quad (3.9)$$

When the traffic states move only along a single branch, and given any initial vehicle accumulation n_1 at the beginning of a period from t_1 to t_2, the number of vehicles n_2 at the end of this period can be determined as:

$$\int_{t_1}^{t_2} dt = -\int_{n_1}^{n_2} \frac{1}{a_i n + b_i - q} dn \qquad (3.10)$$

$$t_2 - t_1 = -\frac{1}{a_i} \ln\left(\frac{a_i n_2 + b_i - q}{a_i n_1 + b_i - q}\right) \qquad (3.11)$$

$$n_2 = \frac{e^{-a_i(t_2-t_1)}(a_i n_1 + b_i - q)}{a_i} - \frac{b_i - q}{a_i} \qquad (3.12)$$

Therefore, with the disruption accumulation n', and when the traffic states are on the same branch. After any time t, the vehicle accumulation n would be:

$$n = \frac{a_1 n' + b_1 - q}{a_1} e^{-a_1 t} - \frac{b_1 - q}{a_1} \qquad (3.13)$$

The TTS in this case can be calculated as:

$$TTS = \int_0^t n\, dt = \int_0^t \left(\frac{a_1 n' + b_1 - q}{a_1} e^{-a_1 t} - \frac{b_1 - q}{a_1}\right) dt \qquad (3.14)$$

$$= -\frac{a_1 n' + b_1 - q}{a_1^2} e^{-a_1 t} - \frac{b_1 - q}{a_1} t + \frac{a_1 n' + b_1 - q}{a_1^2} \qquad (3.15)$$

Now we calculate the derivatives of TTS, considering t as any positive constant.

$$\frac{dTTS}{dn'} = \frac{1}{a_1} - \frac{e^{-a_1 t}}{a_1} \qquad (3.16)$$

$$\frac{d^2 TTS}{dn'^2} = 0 \qquad (3.17)$$

The second derivative of TTS is 0, indicating that when the traffic states move only along a single branch, it shows neither fragility nor antifragility.

On the other hand, when the traffic state goes over the critical vehicle accumulation n_c, and since the MoC is a piece-wise function, we calculate the TTS separately on both the more congested and the less congested branches, denoted as TTS_1 and TTS_2. Since the critical time t_c is still unknown, we need to determine t_c first, similar to Eq. (3.11).

$$t_c = -\frac{1}{a_1} \ln\left(\frac{a_1 n_c + b_1 - q}{a_1 n' + b_1 - q}\right) \qquad (3.18)$$

As both $a_1 n_c + b_1$ and $a_2 n_c + b_2$ are equal to m_c, we can rewrite the above Eq. (3.18) as:

3.2 Discussing the Fragility of Road Transport Systems

$$t_c = -\frac{1}{a_1} \ln\left(\frac{m_c - q}{a_1 n' + b_1 - q}\right) \quad (3.19)$$

Likewise to Eq. (3.15), the TTS_1 for the more congested branch is:

$$TTS_1 = -\frac{a_1 n' + b_1 - q}{a_1^2} e^{-a_1 t_c} - \frac{b_1 - q}{a_1} t_c + \frac{a_1 n' + b_1 - q}{a_1^2} \quad (3.20)$$

$$= -\frac{m_c - q}{a_1^2} + \frac{b_1 - q}{a_1^2} \ln\left(\frac{m_c - q}{a_1 n' + b_1 - q}\right) + \frac{a_1 n' + b_1 - q}{a_1^2} \quad (3.21)$$

Since TTS is the sum of TTS_1 and TTS_2, the second derivative of TTS would also be the sum of the derivatives. The derivatives for TTS_1 are:

$$\frac{dTTS_1}{dn'} = -\frac{b_1 - q}{a_1}(a_1 n' + b_1 - q)^{-1} + \frac{1}{a_1} \quad (3.22)$$

$$\frac{d^2 TTS_1}{dn'^2} = (b_1 - q)(a_1 n' + b_1 - q)^{-2} \quad (3.23)$$

According to Eq. (3.12), the vehicle accumulation on the less congested branch from t_c to t would be:

$$n = \frac{e^{-a_2(t-t_c)}(a_2 n_c + b_2 - q)}{a_2} - \frac{b_2 - q}{a_2} \quad (3.24)$$

$$= \frac{e^{a_2 t_c}(m_c - q)}{a_2} e^{-a_2 t} - \frac{b_2 - q}{a_2} \quad (3.25)$$

The TTS_2 for the less congested branch would be:

$$TTS_2 = \int_{t_c}^{t} n \, dt = \int_{t_c}^{t} \left(\frac{e^{a_2 t_c}(m_c - q)}{a_2} e^{-a_2 t} - \frac{b_2 - q}{a_2}\right) dt \quad (3.26)$$

$$= -\frac{e^{-a_2 t}(m_c - q)}{a_2^2} e^{a_2 t_c} + \frac{b_2 - q}{a_2} t_c + \frac{m_c - q}{a_2^2} - \frac{b_2 - q}{a_2} t \quad (3.27)$$

The derivatives on the less congested branch are:

$$\frac{dTTS_2}{dn'} = -\frac{(m_c - q)^{1-\frac{a_2}{a_1}} e^{-a_2 t}}{a_2}(a_1 n' + b_1 - q)^{\frac{a_2}{a_1} - 1} + \frac{b_2 - q}{a_2}(a_1 n' + b_1 - q)^{-1} \quad (3.28)$$

$$\frac{d^2 TTS_2}{dn'^2}$$
$$= -\left(\frac{(a_2 - a_1)(m_c - q)^{1-\frac{a_2}{a_1}} e^{-a_2 t}}{a_2}(a_1 n' + b_1 - q)^{\frac{a_2}{a_1}} + \frac{a_1(b_2 - q)}{a_2}\right)(a_1 n' + b_1 - q)^{-2}$$
$$(3.29)$$

The second derivative of the whole process can be written as:

$$\frac{d^2 TTS}{dn'^2} = \frac{d^2 TTS_1}{dn'^2} + \frac{d^2 TTS_2}{dn'^2} \qquad (3.30)$$

$$= \left(b_1 - q - \frac{e^{-a_2 t}}{a_2}(a_2 - a_1)(m_c - q)^{1-\frac{a_2}{a_1}}(m_0 - q)^{\frac{a_2}{a_1}} - \frac{a_1(b_2 - q)}{a_2}\right)(m_0 - q)^{-2} \qquad (3.31)$$

As per Assumption 3, we have $m_0 - q > 0$, so if a transportation system is to be fragile, $d^2 TTS/dn'^2$ should also be positive, and the following equation has to be true:

$$b_1 - q - \frac{e^{-a_2 t}}{a_2}(a_2 - a_1)(m_c - q)^{1-\frac{a_2}{a_1}}(m_0 - q)^{\frac{a_2}{a_1}} - \frac{a_1(b_2 - q)}{a_2} > 0 \qquad (3.32)$$

Since $t > t_c$ and $a_2 > a_1$, regardless of whether a_2 is positive or negative, the following relationship always holds:

$$-\frac{e^{-a_2 t}}{a_2} > -\frac{e^{-a_2 t_c}}{a_2} \qquad (3.33)$$

As the following three terms, $a_2 - a_1$, $(m_c - q)^{1-\frac{a_2}{a_1}}$, and $(m_0 - q)^{\frac{a_2}{a_1}}$ are all positive, the following relationship is true:

$$b_1 - q - \frac{e^{-a_2 t}}{a_2}(a_2 - a_1)(m_c - q)^{1-\frac{a_2}{a_1}}(m_0 - q)^{\frac{a_2}{a_1}} - \frac{a_1(b_2 - q)}{a_2} > \qquad (3.34)$$

$$b_1 - q - \frac{e^{-a_2 t_c}}{a_2}(a_2 - a_1)(m_c - q)^{1-\frac{a_2}{a_1}}(m_0 - q)^{\frac{a_2}{a_1}} - \frac{a_1(b_2 - q)}{a_2} \qquad (3.35)$$

Here we substitute t_c in Eq. (3.35) with Eq. (3.19) and we get:

$$b_1 - q - \frac{(a_2 - a_1)(m_c - q)}{a_2} - \frac{a_1(b_2 - q)}{a_2} = \qquad (3.36)$$

$$a_1\left(\frac{b_1 - m_c}{a_1} - \frac{b_2 - m_c}{a_2}\right) = a_1(n_c - n_c) = 0 \qquad (3.37)$$

Hence, we have:

$$b_1 - q - \frac{e^{-a_2 t}}{a_2}(a_2 - a_1)(m_c - q)^{1-\frac{a_2}{a_1}}(m_0 - q)^{\frac{a_2}{a_1}} - \frac{a_1(b_2 - q)}{a_2} > 0 \qquad (3.38)$$

The second derivative of TTS over the disruption vehicle accumulation n' is positive, which indicates the fragility. □

3.2.3.2 Supply Disruption

A positive second derivative of time spent concerning the magnitude of MFD disruption would demonstrate the fragility from the perspective of supply disruptions. Here, we use a supply disruption magnitude coefficient r, and the disrupted MFD is expressed as $(1-r)G(n)$. Although real-world MFD may be decreased in various shapes, we use this simple approach as applied in [35] when studying the uncertainties of MFDs. The physical meaning of $(1-r)G(n)$ relates to the decrease of the free-flow speed due to, e.g., snowy weather and icy roads, with the maximal density of the network remaining unchanged. The traffic demand at equilibrium before the MFD disruption is $q = G(k_0)$, or $q = G(n_0)$. After the supply disruption, as per Assumption 3, the supply disruption magnitude coefficient $r \in [0, 1)$ is not significantly large so the traffic demand remains below the maximal capacity on the disrupted MFD profile, and the new equilibrium point is $q = (1-r)G(k'(r))$, or $q = (1-r)G(n'(r))$. It should be noted that, unlike the study of demand disruption, when studying supply disruptions, $k'(r)$ is a dependent variable on r.

Proposition 3 *Road transportation systems are fragile with the onset of supply disruptions on the macroscopic level.* □

Proof The traffic demand q is constant, and therefore, the traffic density of the new stable state $k'(r)$ would be:

$$q = (1-r)(uk'(r) + c) \tag{3.39}$$

$$k'(r) = \frac{\frac{q}{1-r} - c}{u} \tag{3.40}$$

The ATS and its derivatives are:

$$ATS = \frac{Lk'(r)}{q} = \frac{L}{qu}\left(\frac{q}{1-r} - c\right) \tag{3.41}$$

$$\frac{dATS}{dr} = \frac{L}{u}(1-r)^{-2} \tag{3.42}$$

$$\frac{d^2ATS}{dr^2} = \frac{2L}{u}(1-r)^{-3} \tag{3.43}$$

As per Assumption 2, when studying supply disruptions, we focus on the uncongested zone of the MFD, meaning the slope of these relevant cuts is positive so that both derivatives are positive. □

Proposition 4 *Road transportation systems are fragile when going through the recovery process from supply disruptions.* □

Proof Here, we need to combine the conclusions from Propositions 2 and 3. In Proposition 2, we've proven the recovery process to be fragile when the traffic state

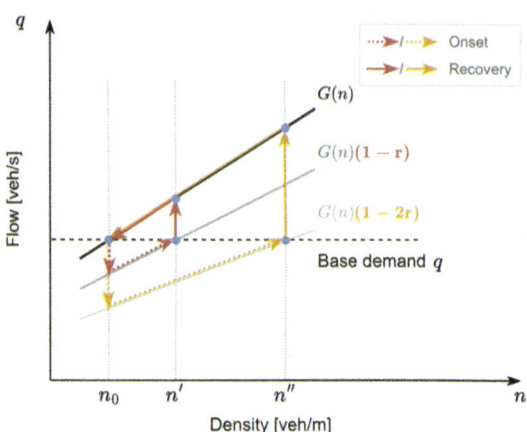

Fig. 3.9 Recovery from supply disruption

shifts from the more congested branch to the less congested branch for any two consecutive cuts on the Daganzo MFD, or to be neither fragile nor antifragile when it stays only on one single branch. Therefore, it can be mathematically summarized as $d^2TTS/dn'^2 \geq 0$. Following Assumption 2, $a_1 > 0$ holds as the branch is below the critical density, we can easily prove the first derivative $dTTS/dn'$ to be non-negative as well with Eqs. (3.22) and (3.28) through the same procedure as proving the second derivative to be positive. And for the proof of recovery from supply disruptions, as shown in Fig. 3.9, likewise to Proposition 3, the relationship between the original and new equilibrium points before and after MFD disruptions can be expressed as:

$$q = un_0 + c = (1-r)(un' + c) \tag{3.44}$$

$$n'(r) = \frac{un_0 + c}{u(1-r)} - c/u \tag{3.45}$$

The first and second derivatives of n' over the supply disruption magnitude coefficient r are:

$$\frac{dn'}{dr} = \frac{un_0 + c}{u}(1-r)^{-2} \tag{3.46}$$

$$\frac{d^2n'}{dr^2} = \frac{2(un_0 + c)}{u}(1-r)^{-3} \tag{3.47}$$

Since u and $un_0 + c$ are both positive, the derivatives of n' over r are positive as well. Additionally, when considering the transition toward a more congested branch, the same conclusion also holds. As TTS is a function of n' and n' is again a function of r, by applying the chain rule, we can get the second derivative of TTS over r as:

$$\frac{d^2 TTS}{dr^2} = \frac{d}{dr}\left(\frac{dTTS}{dn'} \cdot \frac{dn'}{dr}\right) \tag{3.48}$$

$$= \frac{d^2 TTS}{dn_0'^2} \cdot \left(\frac{dn'}{dr}\right)^2 + \frac{dTTS}{dn'} \cdot \frac{d^2 n'}{dr^2} \tag{3.49}$$

Because all the four components of the Eq. (3.49) have been demonstrated above to be non-negative, thus $d^2 TTS/dr^2$ is also non-negative. □

3.3 Inherited Antifragility in Traffic Management

Perimeter control is a well-established traffic management strategy that has been applied in numerous cities worldwide with positive results. Essentially, there is an area that is very important for the network users, and it should remain operational and efficient, like a city center. To protect this area, the traffic signals on its perimeter regulate the inflow (users that want to enter this area) in a way that the traffic state does not become over-saturated, thus leading to undesired results such as gridlock with exponentially increasing delays for the users [33, 36]. The stochastic behavior of network users, their uneven distribution within the area (MFD heterogeneity), the lack of observability through sensors, adverse weather conditions and traffic incidents can alter the shape of the MFDs and potentially violate the mathematical model that serves as the foundation for the established model-based perimeter controllers.

3.3.1 Reinforcement Learning in Traffic Operations

Tackling the parameter uncertainties in the models caused by real-world disruptions, non-parametric learning-based approaches in traffic control seem promising [37]. Reinforcement learning (RL) has been researched extensively in transportation operations, and recent works [31, 38] have illustrated the efficiency of RL algorithms with the benefit of the exact system dynamics being agnostic.

Despite the popularity of RL-based control, only a couple of studies focus on resilience and assess their proposed method with scenarios under traffic demand uncertainties and MFD stochasticity [39, 40]. When examining the system's robustness, the types of scenarios under consideration become more diverse, including incidents and sensor failures [41–43].

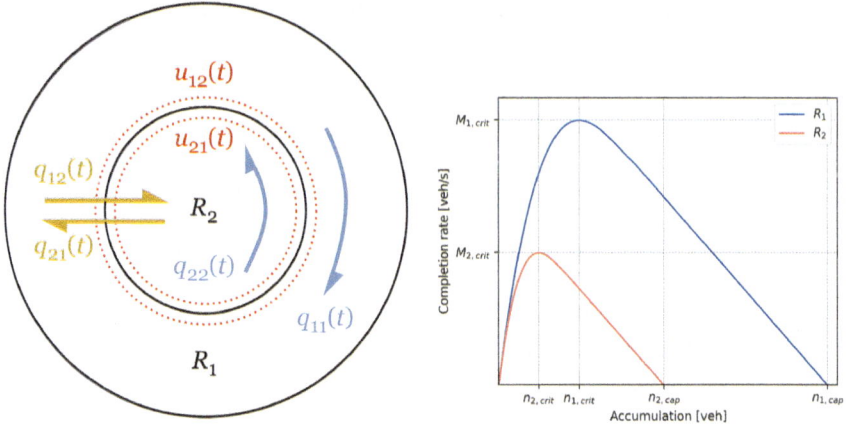

(a) The cordon-shaped urban network (b) MFDs for the inner and outer regions

Fig. 3.10 The network structure and the related MFDs

3.3.2 Antifragile Perimeter Control

We study the problem of perimeter control between two homogeneous regions. A cordon-shaped urban network is investigated, with the inner region representing a city center, as shown in Fig. 3.10a. Traffic demand for an Origin-Destination (OD) pair from region i to region j at time t is denoted as $q_{ij}(t)$. The inner and outer regions have different MFDs due to the difference in capacity to accommodate vehicles in the road networks within the city center and the surrounding region, defined as $G_i(n_i(t))$ as illustrated in Fig. 3.10b. Given the total number of vehicles with presence in region i at time t, denoted as $n_i(t)$, the total trip completion rate for this region i, denoted as $M_i(t)$, can be determined using the corresponding MFD, which comprises both the intra-regional trip completion, i.e., $M_{ii}(t)$ and the inter-regional transfer flow, i.e., $M_{ij}(t)$ ($i \neq j$) with $i, j \in \{1, 2\}$. To protect both regions from being overflown by possible high traffic demand, the percentage of the transfer flow allowed to go across the region perimeter at time t is regulated by two perimeter controllers denoted as $u_{ij}(t)$ ($i \neq j$). A list of all notations used in this paper, including the notations used in defining the RL algorithm and the antifragile terms, is summarized in Table 3.1.

In this context, Eq. (3.50) describes the change rate of the intra-regional vehicle accumulation of region i. It is the sum of intra-regional traffic demand in this region, denoted as $q_{ii}(t)$, together with the perimeter control regulated transfer flow from region j to region i, defined as $u_{ji}(t) \cdot M_{ji}(t)$, then deducted by the trip completion within region i, denoted as $M_{ii}(t)$. Likewise, the change rate of inter-regional traffic accumulation, as in Eq. (3.51) shows, is the difference between the inter-regional traffic demand, denoted as $q_{ij}(t)$, and the regulated transfer flow $u_{ij}(t) \cdot M_{ij}(t)$:

3.3 Inherited Antifragility in Traffic Management

Table 3.1 List of notations

Symbol	Meaning
1. General notations in problem formulation	
t	Time
Δt	Time step
t_{end}	Total simulation time
$n_{ij}(t)$	Vehicle accumulation with OD from region i to j at time t
$n_i(t)$	Vehicle accumulation in region i at time t
$u_{ij}(t)$	Perimeter control variables regulating flow from region i to j at time t
$q_{ij}(t)$	Traffic demand with OD pair i and j at time t
$G_i(n_i(t))$	Sum of trip completion and transfer flow in region i at time t
$M_{ij}(t)$	Trip completion with OD from region i to j ($i \neq j$) at time t
$n_{i,\text{cap}}(t)$	Maximal number of vehicles (jam accumulation) in region i at time t
$n_{i,\text{crit}}(t)$	Vehicle accumulation with highest completion rate in region i at time t
J	Objective function
2. Notations in reinforcement learning	
$\mathcal{S}$	State space, the whole set of states the RL agent can transition to
s_t	$s_t \in \mathcal{S}$, the observable state in simulation at time t
$\mathcal{A}$	Action space, the whole set of actions the RL agent can act out
a_t	$a_t \in \mathcal{A}$, the action taken in simulation at time t
$\mathcal{R}$	The reward function for the RL agent
r_t	$r_t = \mathcal{R}(s_t, a_t)$, the received reward with state s_t and action a_t at time t
γ	Discount factor to favor rewards in the near future
$Q(s(t), a(t))$	Expected long-term return for taking action $a(t)$ in state $s(t)$ at time t

(continued)

Table 3.1 (continued)

Symbol	Meaning
θ^μ	Weight parameter of the deep neural network for the actor-network
θ^Q	Weight parameter of the deep neural network for the critic network
y_i	Expected long-term return calculated with the target critic network
L	The loss of the critic network
ρ^β	All possible trajectories of s_t
I	The objective function for the actor-network
3. Notations for the additional antifragile terms applied in reinforcement learning	
$r_{add}(t)$	Additional reward term in RL based on derivatives and redundancy
ω_h	The weight of first derivative in the additional reward term $r_{add}(t)$
$\omega_{\Delta h}$	The weight of second derivative in the additional reward term $r_{add}(t)$
$\alpha_i(t)$	Binary variable determining the term to be reward/penalty
$h_i(t)$	The first derivative of traffic state at time t
$\Delta h_i(t)$	The second derivative of traffic state at time t

3.3 Inherited Antifragility in Traffic Management

$$\frac{dn_{ii}(t)}{dt} = q_{ii}(t) + u_{ji}(t) \cdot M_{ji}(t) - M_{ii}(t) \tag{3.50}$$

$$\frac{dn_{ij}(t)}{dt} = q_{ij}(t) - u_{ij}(t) \cdot M_{ij}(t), \ (i \neq j) \tag{3.51}$$

The total trip completion, i.e., $M_i(t)$ for region i at time t is calculated based on the trip accumulation and the related MFD, defined as $G_i(n_i(t))$, and is the sum of the intra-regional trip completion, i.e., $M_{ii}(t)$, in Eq. (3.52) and the inter-regional transfer flow, i.e., $M_{ij}(t)$ ($i \neq j$), in Eq. (3.53):

$$M_{ii}(t) = \frac{n_{ii}(t)}{n_i(t)} \cdot G_i(n_i(t)) \tag{3.52}$$

$$M_{ij}(t) = \frac{n_{ij}(t)}{n_i(t)} \cdot G_i(n_i(t)), \ (i \neq j) \tag{3.53}$$

$$n_i(t) = \sum_{j=1,2} n_{ij}(t) \tag{3.54}$$

The objective function is to maximize the throughput of this cordon-shaped network, which is the sum of the intra-regional trip completion in both regions.

$$J = \max_{u_{ij}(t)} \int_0^{t_{end}} \sum_{i=1,2} M_{ii}(t) dt \tag{3.55}$$

subject to the following boundary conditions:

$$n_{ij}(t) \geq 0 \tag{3.56}$$

$$n_i(t) \leq n_{i,cap} \tag{3.57}$$

$$u_{min} \leq u_{ij}(t) \leq u_{max} \tag{3.58}$$

Intra-regional and inter-regional vehicle accumulation, i.e., $n_{ii}(t)$ and $n_{ij}(t)$ are non-negative values, and $n_{i,cap}$ is the maximal possible number of vehicles accumulated in the region i. At this vehicle accumulation, a gridlock will occur in the network. u_{min} and u_{max} represent the lower and upper limits for the perimeter control variable $u_{ij}(t)$ for both directions, and such applications are in line with [31, 33]. These bounds are since perimeter control is normally implemented through signalization. While u_{max} accounts for the lost time caused by the interchange between the red and green phases, $u_{\min}$ is necessitated since an indefinitely long red light is rare in real-world cases.

In contrast to the control-based strategies, for the RL-based algorithms, following the idea of redundancy, an additional term $r_{add}(t)$ is added into the objective function J_{RL}, referred to as reward $r_t \in \mathcal{R}$ in the context of RL, leading to:

$$J_{RL} = \max_{u_{ij}(t)} \int_0^{t_{end}} [\sum_{i=1,2} M_{ii}(t) + r_{add}(t)]dt \qquad (3.59)$$

The term $r_{add}(t)$ aims to build up a proper redundancy so that the proposed RL algorithm does not reward the agent for targeting the exact critical accumulation point. A comprehensive explanation of the term $r_{add}(t)$ for the reward in RL will be provided in the following section.

3.3.3 Reinforcement Learning Mechanism

In RL algorithms, an agent or multiple agents interact with a preset environment and improve the performance of decision-making, defined as action a_t in an action space $\mathcal{A}$, based on the observable state s_t in the state space $\mathcal{S}$ and the reward, defined as $r_t = \mathcal{R}(s_t, a_t)$, where $\mathcal{R}$ is the reward function. The improvement of decision-making is commonly realized through a deep neural network as a function approximation.

The RL algorithm applied in this work is Deep Deterministic Policy Gradient (DDPG), as proposed in [44]. By applying an actor-critic scheme, DDPG can manage a continuous action space instead of only choosing from a limited set of discrete values as in the Deep Q-Network (DQN) algorithm [45], which are commonly applied as Table 3.1 shows. Also, [31] has demonstrated that an RL algorithm with a continuous action space can achieve better performance compared to a discrete action space. The DDPG algorithm can be divided into two main components, namely the actor and the critic, which are updated at each step through policy gradient and Q-value, respectively. The scheme of the DDPG algorithm applied in this work is schematically illustrated in Fig. 3.11.

The state $s_t \in \mathcal{S}$ is defined distinctively according to different methods applied in this work. Our proposed method consists of three terms, the vehicle accumulation regarding the OD pair $n_{ij}(t)$, the change rate of vehicle accumulation at each time step $dn_{ij}(t)$ (first derivative) as well as the second derivative $d^2 n_{ij,t}$. In [31], a state s_t defined as $[n_{ij}(t), q_{ij}(t)]$ is adopted. However, since traffic demand in the real world is hardly measurable, $q_{ij}(t)$ would be an unobservable state for the agent. The action $a_t \in \mathcal{A}$ is defined the same as the control variables $u_{ij}(t)$. For the reward r_t, while [31] uses merely the completion rate, in our proposed method, the reward is defined with an additional $r_{add}(t)$ term, as Eq. (3.59) shows.

3.3 Inherited Antifragility in Traffic Management

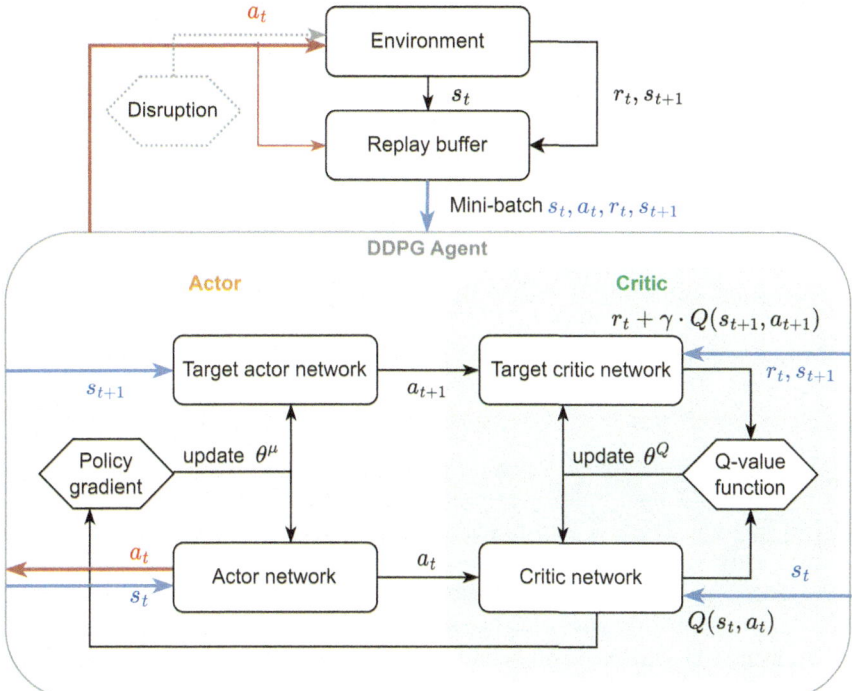

Fig. 3.11 DDPG scheme

The actor-network is represented by $\mu(\cdot)$, and it determines the best action a_t, which is the percentage of vehicles that are allowed to travel across the periphery, based on the current state s_t and the weight parameters at a certain time step t.

The nature of the best action a_t was also explored in the previous work of [46], where, in the framework of variable structure control, the authors demonstrate the need for a discontinuous signal. The critic network, denoted by $Q(\cdot)$, takes the responsibility to evaluate whether a specific state-action pair at a certain time step yields the maximal possible discounted future reward $Q(s_t, a_t)$. A common technique used in DDPG is to create a target actor network $\mu'(\cdot)$ and a target critic network $Q'(\cdot)$, which are a copy of the original actor and critic network but updated posteriorly to stabilize the training process and prevent overfitting [47]. The target maximal discounted future reward for the target critic network can be calculated as in Eq. (3.60).

$$y_i = r_i + \gamma Q'(s_{i+1}, \mu'(s_{i+1}|\theta^{\mu'})|\theta^{Q'}) \tag{3.60}$$

Similar to DQN, the critic network can be updated by calculating the temporal difference between the predicted reward and the target reward and minimizing the loss for a mini-batch N sampled from the replay buffer:

$$L = \frac{1}{N} \sum_i (y_i - Q(s_i, a_i | \theta^Q))^2 \tag{3.61}$$

Afterwards, the actor-network can be updated with sampled deterministic policy gradient [44]:

$$J = \mathbb{E}_{s_t \sim \rho^\beta}[r(s, \mu(s|\theta^\mu))|_{s=s_t}] \tag{3.62}$$

$$\nabla_{\theta^\mu} J = \mathbb{E}_{s_t \sim \rho^\beta}[\nabla_a Q(s, a|\theta^Q)|_{s=s_t, a=\mu(s_t)} \nabla_{\theta_\mu} \mu(s|\theta^\mu)|_{s=s_t}] \tag{3.63}$$

$$\nabla_{\theta^\mu} J \approx \frac{1}{N} \sum_t \nabla_a Q(s, a|\theta^Q)|_{s=s_t, a=\mu(s_t)} \nabla_{\theta^\mu} \mu(s|\theta^\mu)|_{s_t} \tag{3.64}$$

When training the agent, we add disruptions into the simulation environment starting from a certain training episode in the form of surging traffic demand or MFD disruption. These various forms of volatility ought to optimally elicit various learning and decision-making processes. The RL agent would include the results of their prior choices to create and update their weight parameters in a situation with high outcome volatility and being capable of generating and updating expectations after sensing a change in a high-volatility environment.

3.3.4 Antifragile Reinforcement Learning

We realize antifragile behavior by modifying the state space $\mathcal{S}$. We add additional terms based on derivatives and redundancy [23, 48] in both the state space $\mathcal{S}$ and the reward function $\mathcal{R}$ of the RL algorithm.

For the state space $\mathcal{S}$, we replace q_{ij} using the first and second derivatives of the vehicle accumulation $dn_{ij}(t)$ and $d^2 n_{ij}(t)$. With this additional information, the RL agent is aware of a possible demand surge of MFD disruption, and $d^2 n_{ij}(t)$ can reflect the curvature of such changes.

For the reward function $\mathcal{R}$, the trip completion at each time step acts as the main component of our proposed method. The term $r_{add}(t)$ in the objective function J_{RL} in Eq. (3.59) acts as an additional term to build up redundancy in the system. Similar to the creation of the additional term in the state space, we create redundancy also through the calculation of the derivatives, but instead of the derivatives of the vehicle accumulation, we calculate the derivatives of the traffic state. This creates a one-to-one correspondence between the derivatives of the vehicle accumulation and the derivatives of the traffic state. To explain this antifragile term in the reward, we summarize $r_{add}(t)$ as the sum of two terms, with $H(t)$ being an overall term representing the first derivative of the traffic state and $\Delta H(t)$ representing the second derivative:

$$r_{add}(t) = H(t) + \Delta H(t) \tag{3.65}$$

3.3 Inherited Antifragility in Traffic Management

Here, $H(t)$ and $\Delta H(t)$ can be expanded as:

$$H(t) = \sum_{i=1,2} H_i(t) = \omega_h \sum_{i=1,2} f(n_i(t), n_{i,\text{crit}}, n_{i,\text{cap}}) \cdot \alpha_i(t) \cdot h_i(t) \quad (3.66)$$

$$\Delta H(t) = \sum_{i=1,2} \Delta H_i(t) = \omega_{\Delta h} \sum_{i=1,2} f(n_i(t), n_{i,\text{crit}}, n_{i,\text{cap}}) \cdot \Delta h_i(t) \quad (3.67)$$

$h_i(t)$ and $\Delta h_i(t)$ are the first and second numerical derivatives of the traffic states on the MFD, $h_i(t)$ is defined as the difference of trip completion over vehicle accumulation at the end of a time step versus at the beginning of the same time step, as in Eq. (3.68) shows, and the second derivative $\Delta h_i(t)$ is calculated as the difference between the first derivatives of two consecutive time steps, as in Eq. (3.69) shows:

$$h_i(t) = \frac{M_i(t) - M_i(t-1)}{n_i(t) - n_i(t-1)} \quad (3.68)$$

$$\Delta h_i(t) = h_i(t) - h_i(t-1) \quad (3.69)$$

All variables involved in the deep neural network should be normalized to facilitate the training process, meaning the exact values of the derivatives are not of importance. Hence ω_h and $\omega_{\Delta h}$ are introduced as the weight constants for the first and second derivatives to regulate their impact on the reward side $\mathcal{R}$.

The binary variable $\alpha_i(t)$ was designed in the first derivative to reward the agent when moving towards the desired direction on the MFD. For instance, the derivative of any data point in the congested zone of the MFD is negative. In this case, when the vehicle accumulation is still getting larger, a penalty will be applied. However, if the vehicle accumulation is decreasing through perimeter control, this binary $\alpha_i(t)$ variable will turn it into a reward. For the second derivative, an additional binary variable is not necessary since the two consecutive first derivatives can determine whether $\Delta h_i(t)$ is either positive or negative.

$$\alpha_i(t) = \begin{cases} 1, & \text{if } n_i(t) \geq n_i(t-1), \\ -1, & \text{otherwise.} \end{cases} \quad (3.70)$$

The term $f(n_i(t), n_{i,\text{crit}}, n_{i,\text{cap}})$ is a reduction factor to constrain the impact of the $r_{add}(t)$ term when the accumulation is either on a much lower level (empty network) or on a much higher level (gridlock). The area near the critical accumulation is where the $r_{add}(t)$ term should have the greatest impact. Here we use a modified trigonometric function to realize this purpose. It should be noted that other functions, such as normal distribution, are also valid for achieving the same purpose.

$$f(n_i(t), n_{i,\text{crit}}, n_{i,\text{cap}}) = \begin{cases} \dfrac{1 + \cos\left(-\pi \cdot \dfrac{n_{i,\text{crit}} - n_i(t)}{n_{i,\text{crit}}}\right)}{2}, & \text{if } n_i(t) \geq n_{i,\text{crit}}, \\ \dfrac{1 + \cos\left(-\pi \cdot \dfrac{n_i(t) - n_{i,\text{crit}}}{n_{i,\text{cap}} - n_{i,\text{crit}}}\right)}{2}, & \text{otherwise.} \end{cases}$$

(3.71)

After considering all the modifiers above, we show $H(t)$ and $\Delta H(t)$ using a single MFD as an example in Figs. 3.12 and 3.13. The first derivative $H(t)$, in Fig. 3.12a, rewards the agent more when it's moving towards the critical accumulation to maximize its trip completion rate. However, when the number of vehicles approaches the critical accumulation, this term drops significantly and becomes a penalty when it exceeds the critical point. Since the first derivative $H(t)$ is a complementary term in addition to the trip completion in the reward function, we showcase the influence of this term on the MFD after normalization in Fig. 3.12. With increasing weight coefficient ω_h, the critical accumulation of the modified MFD becomes marginally smaller compared to the original MFD, and the reward that the RL agent can receive also decreases faster after the accumulation exceeds the critical accumulation. Although the trip completion still follows the original MFD in the simulation environment. The agent learns to get more rewards following the modified MFD. In this way, redundant overcompensation has been established to prevent accumulation from exceeding the critical accumulation when disruption takes place unexpectedly.

An interesting note is that estimation uncertainty, also known as second-order uncertainty, is another factor that affects disruptions that take place unexpectedly. This is the imprecision of the learner's current beliefs about the environment, and what the antifragile terms capture. This amount reduces with sampling if beliefs are acquired by learning as opposed to instruction (e.g. anticipation through redundant overcompensation). When estimating uncertainty is substantial, unlikely samples could partly be attributed to the agent's false assumptions about the environment's structure rather than a change in that structure (e.g. around critical accumulation).

The second derivative $\Delta H(t)$ is shown in Fig. 3.13. The x-axis is the vehicle accumulation, same as in Fig. 3.12a, while the y-axis represents how fast the traffic state is changing on the MFD. The faster it increases to reach the critical accumulation, the greater the penalty will be applied to the RL agent. This observation is consistent with the redundant overcompensation and time-scale separation principles formalized in [23] and practically applied in [49, 50]. On the contrary, if the vehicle accumulation decelerates, a reward will be applied. Similar to $H(t)$, this complementary term $\Delta H(t)$ is also dependent on the normalization factor $\omega_{\Delta h}$.

With $H(t)$ and $\Delta H(t)$, the agent learns to be conservative when regulating the perimeter control variables when the accumulation is about to reach critical, in case disruptions take place. Therefore, as can be concluded, although $H(t)$ and $\Delta H(t)$ apply the same concept of the derivative as the $dn_{ij}(t)$ and d^2n_{ij} in the state space $\mathcal{S}$, the purpose of $H(t)$ and $\Delta H(t)$ is preserving redundancy in the system instead of feeding additional information to the agent. This behavior is consistent with the

3.3 Inherited Antifragility in Traffic Management   37

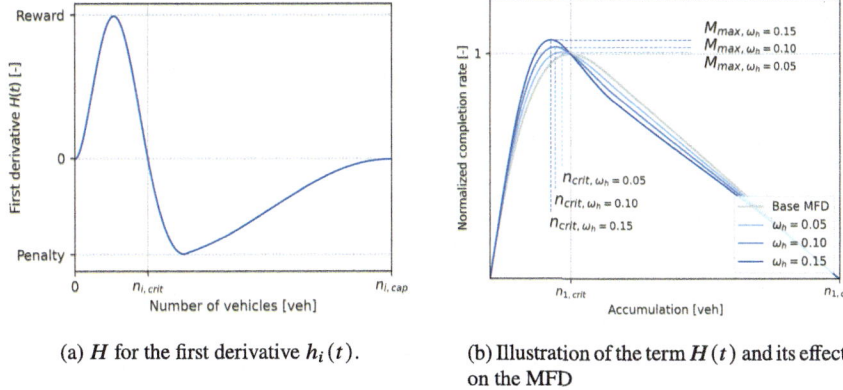

(a) H for the first derivative $h_i(t)$.

(b) Illustration of the term $H(t)$ and its effect on the MFD

Fig. 3.12 Illustration of the term $H(t)$ and its effect on the MFD

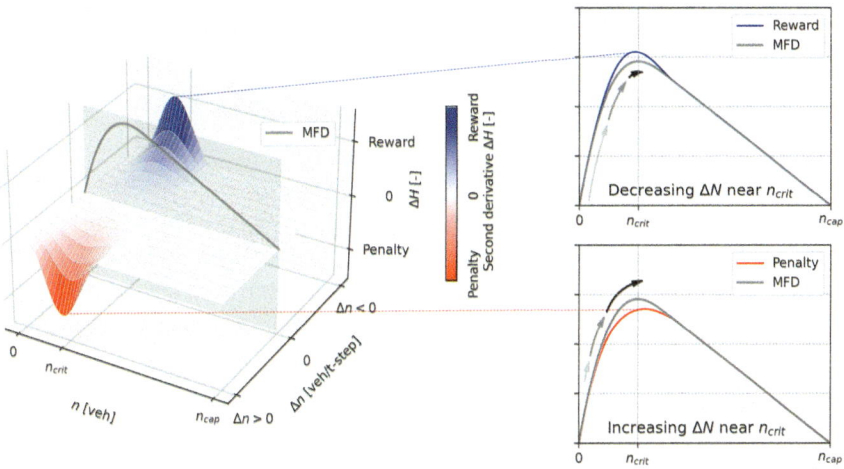

Fig. 3.13 Illustration of the term $\Delta H(t)$

locally discontinuous shape of the action signals $u_{ij}(t)$ applied to the cordon network, as suggested by the control theoretic study of [46].

3.3.5 Results Showing Inherited Antifragility

A simulation was used to evaluate the antifragile design. In the corresponding study [51], three perimeter control strategies in addition to a no-control scenario are compared.

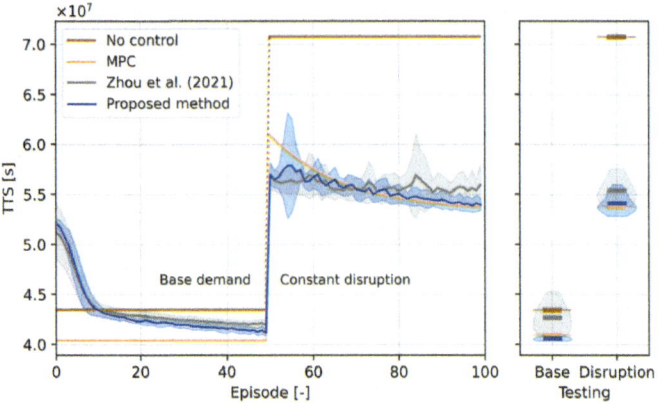

Fig. 3.14 Performance curves under constant disruptions

- No control.
- MPC modified from [33], averaging all history profiles including disruption as a new demand or MFD profile.
- State-of-the-art RL-based method proposed by [31] as baseline:
  State s_t: vehicle accumulation and traffic demand $[n_{ij}(t), q_{ij}(t)]$.
  Reward r_t: trip completion in both regions $[\sum_{i=1,2} M_{ii}(t)]$.
- The proposed RL-based antifragile method:
  State s_t: vehicle accumulation and its derivatives $[n_{ij}(t), dn_{ij}(t), d^2 n_{ij}(t)]$.
  Reward r_t: trip completion and the $r_{add}(t)$ redundancy term $[\sum_{i=1,2} M_{ii}(t) + r_{add}(t)]$.

Here we will briefly discuss the results based on proto-antifragility and for more insights, we refer the reader to the corresponding publication [51]. Systems with the property of proto-antifragility can learn from past adverse events and anticipate possible ongoing disruptions of similar magnitudes to enhance future performance. To validate this property, we apply disruptions with constant magnitude after training the agent with the base demand profile for a certain number of episodes.

Unlike the no-control and MPC approaches, RL-based methods are subject to performance variations over simulation episodes. Therefore, in Fig. 3.14, besides the performance curves of the no-control and MPC approaches, we also show the learning curves of the baseline RL method and our proposed antifragile RL method. The curves are averaged from 15 simulations, while the shadowed area indicates the standard deviation of the simulation results. After conducting 50 episodes of training together with testing under disruptions, the performance of different algorithms under no disruptions is then again validated. Hence, an addition of 2 testing phases was implemented under and under no disruption, using the weight parameters of the neural networks obtained from the last training episode.

It can be noticed that two significant phenomena occurred during the initial 50 episodes with the base demand. First, the proposed antifragile method can achieve better performance with less TTS when there is no disruption in the network. While the baseline method has almost converged after 50 episodes, the proposed antifragile method seems to be capable of further improving the performance with more training episodes. The reason for not continuing the training process until the convergence of both methods is to avoid overtraining the non-disruption-related weight parameters as well as the underfitting of the disruption-related parameters. In addition, the shadowed area of our proposed method represents a less significant performance variation of about 6.0% less compared to the baseline method, based on the averaged performance between episodes 30 and 50. This indicates that the proposed antifragile method exhibits comparably higher training stability.

After disruption with constant magnitude is introduced into the network. The performance of the two RL-based methods has first changed drastically. Then, while the proposed method has an obvious tendency to reduce TTS over the following 50 episodes, the baseline method shows little sign of improving performance. Also, the training stability of the proposed method becomes increasingly stable over episodes as the width of the shadowed area gets smaller.

While the results of testing under disruption follow the same pattern as the last episode of the training process, however, when the simulation environment reverts to the previous no-disruption condition, the performance of different methods behaves differently. Since our modified MPC averages demand history, after experiencing episodes with disruptions, performance deterioration of 1.3% can be observed when there is no disruption. A similar pattern can be observed for the baseline RL method. Not only is the TTS 1.5% higher in this testing episode compared to the training episode 50, but the performance variation is also significantly larger. On the contrary, our proposed antifragile RL method can both maintain excellent performance and achieve smaller variation.

3.3.6 Conclusions

In this chapter, we discussed inherited antifragility in technical complex systems. Fragility in the context of complex systems, refers to their susceptibility to disruptions. The road transportation system is an example of a complex fragile system that can collapse under disruptions and lead to dramatic delays in the travel times of its users. Traffic management attempts to retain the system operation and around an optimal point near the capacity of the system under heavy demand. We mathematically analyzed the fragility of such systems based on empirically derived models like the Macroscopic Fundamental Diagram and finally, we demonstrated how antifragile design in traffic management can learn from volatility and help solutions to respond better in future disruptions, thus demonstrating inherited antifragility.

Acknowledgements We acknowledge the contribution of Mr. Anastasios Kouvelas and Mr. Linghang Sun from ETH Zürich in writing this chapter.

References

1. Laval, J. Self-organized criticality of traffic flow: Implications for congestion management technologies. *Transportation Research Part C: Emerging Technologies*. **149** pp. 104056 (2023)
2. Axenie, C., Scherr, W., Wieder, A., Torres, A., Meng, Z., Du, X., Sottovia, P., Foroni, D., Grossi, M., Bortoli, S. & Others Fuzzy modelling and inference for physics-aware road vehicle driver behaviour model calibration. *Expert Systems With Applications*. **241** pp. 122590 (2024)
3. Makridis, M. & Kouvelas, A. Adaptive physics-informed trajectory reconstruction exploiting driver behavior and car dynamics. *Scientific Reports*. **13**, 1121 (2023)
4. Goodwin, P. Empirical evidence on induced traffic. *Transportation*. **23**, 35–54 (1996,2), https://doi.org/10.1007/BF00166218
5. Fiori, C., Arcidiacono, V., Fontaras, G., Makridis, M., Mattas, K., Marzano, V., Thiel, C. & Ciuffo, B. The effect of electrified mobility on the relationship between traffic conditions and energy consumption. *Transportation Research Part D: Transport And Environment*. **67** pp. 275–290 (2019), Publisher: Pergamon
6. Ni, Y., Makridis, M. & Kouvelas, A. Bicycle as a traffic mode: From microscopic cycling behavior to macroscopic bicycle flow. *Journal Of Cycling And Micromobility Research*. **2** pp. 100022 (2024), https://www.sciencedirect.com/science/article/pii/S2950105924000135, Publisher: Elsevier
7. Makridis, M., Mattas, K., Ciuffo, B. & Kouvelas, A. Impacts of Partially Connected and Automated Vehicles on Traffic Flow and Energy Based on Worldwide Experimental Observations in Motorway Driving. *Sustainable Automated And Connected Transport*. **19** pp. 23–45 (2024,1), https://doi.org/10.1108/S2044-994120240000019002
8. Ballo, L., Freitas, L., Meister, A. & Axhausen, K. The E-Bike City as a radical shift toward zero-emission transport: Sustainable? Equitable? Desirable?. *Journal Of Transport Geography*. **111** pp. 103663 (2023,7), https://www.sciencedirect.com/science/article/pii/S0966692323001357
9. Apostolakis, T., Makridis, M., Kouvelas, A. & Ampountolas, K. Energy-Based Assessment and Driving Behavior of ACC Systems and Humans Inside Platoons. *IEEE Transactions On Intelligent Transportation Systems*. **24**, 12726–12735 (2023,11), https://ieeexplore.ieee.org/abstract/document/10159555, Conference Name: IEEE Transactions on Intelligent Transportation Systems
10. Roads, U. Traffic Assignment Manual for Application with a Large, High Speed Computer. (U.S. Department of Commerce, Bureau of Public Roads, Office of Planning, Urban Planning Division,1964), Google-Books-ID: gkNZAAAAMAAJ
11. Gibson, S., Cooper, G. & Ball, B. The Evolution of Capacity Charges on the UK Rail Network. *Journal Of Transport Economics And Policy*. **36**, 341–354 (2002), https://www.jstor.org/stable/20053906, Publisher: [London School of Economics, University of Bath, London School of Economics and the University of Bath, London School of Economics and Political Science]
12. Corman, F., D'Ariano, A. & Hansen, I. Evaluating Disturbance Robustness of Railway Schedules. *Journal Of Intelligent Transportation Systems*. **18**, 106–120 (2014,1), https://doi.org/10.1080/15472450.2013.801714, Publisher: Taylor & Francis eprint: https://doi.org/10.1080/15472450.2013.801714
13. Iliopoulou, C. & Makridis, M. Critical multi-link disruption identification for public transport networks: A multi-objective optimization framework. *Physica A: Statistical Mechanics And Its Applications*. **626** pp. 129100 (2023,9), https://www.sciencedirect.com/science/article/pii/S0378437123006556
14. Leclercq, L., Ladino, A. & Becarie, C. Enforcing optimal routing through dynamic avoidance maps. *Transportation Research Part B: Methodological*. **149** pp. 118–137 (2021,7), https://www.sciencedirect.com/science/article/pii/S0191261521000813
15. Isaacson, D., Robinson, J., Swenson, H. & Denery, D. A Concept for Robust, High Density Terminal Air Traffic Operations. *10th AIAA Aviation Technology, Integration, And Operations (ATIO) Conference*. (2010), https://arc.aiaa.org/doi/abs/10.2514/6.2010-9292, eprint: https://arc.aiaa.org/doi/pdf/10.2514/6.2010-9292

16. Axenie, C., López-Corona, O., Makridis, M., Akbarzadeh, M., Saveriano, M., Stancu, A. & West, J. Antifragility in complex dynamical systems. *Npj Complexity*. **1**, 12 (2024,8), https://doi.org/10.1038/s44260-024-00014-y
17. Kitchin, R. The Data Revolution: Big Data, Open Data, Data Infrastructures and Their Consequences. (SAGE,2014,9), Google-Books-ID: GfOICwAAQBAJ
18. Duan, Y. & Lu, F. Robustness of city road networks at different granularities. *Physica A: Statistical Mechanics And Its Applications*. **411** pp. 21–34 (2014,10), https://www.sciencedirect.com/science/article/pii/S037843711400466X
19. Shang, W., Gao, Z., Daina, N., Zhang, H., Long, Y., Guo, Z. & Ochieng, W. Benchmark Analysis for Robustness of Multi-Scale Urban Road Networks Under Global Disruptions. *IEEE Transactions On Intelligent Transportation Systems*. **24**, 15344–15354 (2023,12), https://ieeexplore.ieee.org/document/9714769/
20. Mattsson, L. & Jenelius, E. Vulnerability and resilience of transport systems – A discussion of recent research. *Transportation Research Part A: Policy And Practice*. **81** pp. 16–34 (2015,11), https://www.sciencedirect.com/science/article/pii/S0965856415001603
21. Calvert, S. & Snelder, M. A methodology for road traffic resilience analysis and review of related concepts. *Transportmetrica A: Transport Science*. **14**, 130–154 (2018,1), https://doi.org/10.1080/23249935.2017.1363315, Publisher: Taylor & Francis eprint: https://doi.org/10.1080/23249935.2017.1363315
22. Zhou, Y., Wang, J. & Yang, H. Resilience of Transportation Systems: Concepts and Comprehensive Review. *IEEE Transactions On Intelligent Transportation Systems*. **20**, 4262–4276 (2019,12), https://ieeexplore.ieee.org/document/8602445, Conference Name: IEEE Transactions on Intelligent Transportation Systems
23. Taleb, N. & Douady, R. Mathematical definition, mapping, and detection of (anti)fragility. *Quantitative Finance*. **13**, 1677–1689 (2013,11), https://doi.org/10.1080/14697688.2013.800219, Publisher: Routledge eprint: https://doi.org/10.1080/14697688.2013.800219
24. Taleb, N. Antifragile: Things That Gain from Disorder. (Random House Publishing Group,2012,11), Google-Books-ID: 5fqbzqGi0AC
25. Jensen, J. Sur les fonctions convexes et les inégalités entre les valeurs moyennes. *Acta Mathematica*. **30**, 175–193 (1906,1), https://projecteuclid.org/journals/acta-mathematica/volume-30/issue-none/Sur-les-fonctions-convexes-et-les-in
26. Ruel, J., Ayres, M., Ruel, J. & Ayres, M. Jensen's inequality predicts effects of environmental variation. *Trends In Ecology & Evolution*. **14**, 361–366 (1999,9), https://www.cell.com/trends/ecology-evolution/abstract/S0169-5347(99)01664-X, Publisher: Elsevier
27. Ng, M. & Waller, S. A computationally efficient methodology to characterize travel time reliability using the fast Fourier transform. *Transportation Research Part B: Methodological*. **44**, 1202–1219 (2010,12), https://www.sciencedirect.com/science/article/pii/S0191261510000238
28. Greenshields, B., Thompson, J., Dickinson, H. & Swinton, R. The Photographic method of studying traffic behavior. *Highway Research Board Proceedings*. **13** (1934), https://trid.trb.org/View/120821
29. Daganzo, C. A variational formulation of kinematic waves: basic theory and complex boundary conditions. *Transportation Research Part B: Methodological*. **39**, 187–196 (2005,2), https://www.sciencedirect.com/science/article/pii/S0191261504000487
30. Haddad, J. & Shraiber, A. Robust perimeter control design for an urban region. *Transportation Research Part B: Methodological*. **68** pp. 315–332 (2014,10), https://www.sciencedirect.com/science/article/pii/S0191261514001179
31. Zhou, D. & Gayah, V. Model-free perimeter metering control for two-region urban networks using deep reinforcement learning. *Transportation Research Part C: Emerging Technologies*. **124** pp. 102949 (2021,3), https://www.sciencedirect.com/science/article/pii/S0968090X20308269
32. Kouvelas, A., Saeedmanesh, M. & Geroliminis, N. Enhancing model-based feedback perimeter control with data-driven online adaptive optimization. *Transportation Research Part B: Methodological*. **96** pp. 26–45 (2017,2), https://www.sciencedirect.com/science/article/pii/S019126151630710X

33. Geroliminis, N., Haddad, J. & Ramezani, M. Optimal Perimeter Control for Two Urban Regions With Macroscopic Fundamental Diagrams: A Model Predictive Approach. *IEEE Transactions On Intelligent Transportation Systems.* **14**, 348–359 (2013,3), https://ieeexplore.ieee.org/document/6353591, Conference Name: IEEE Transactions on Intelligent Transportation Systems
34. Daganzo, C. & Geroliminis, N. An analytical approximation for the macroscopic fundamental diagram of urban traffic. *Transportation Research Part B: Methodological.* **42**, 771–781 (2008,11), https://www.sciencedirect.com/science/article/pii/S0191261508000799
35. Ambühl, L., Loder, A., Bliemer, M., Menendez, M. & Axhausen, K. A functional form with a physical meaning for the macroscopic fundamental diagram. *Transportation Research Part B: Methodological.* **137** pp. 119–132 (2020,7), https://www.sciencedirect.com/science/article/pii/S0191261517310123
36. Keyvan-Ekbatani, M., Kouvelas, A., Papamichail, I. & Papageorgiou, M. Exploiting the fundamental diagram of urban networks for feedback-based gating. *Transportation Research Part B: Methodological.* **46**, 1393–1403 (2012,12), https://www.sciencedirect.com/science/article/pii/S0191261512000926
37. Nguyen, H., Kieu, L., Wen, T. & Cai, C. Deep learning methods in transportation domain: a review. *IET Intelligent Transport Systems.* **12**, 998–1004 (2018), https://onlinelibrary.wiley.com/doi/abs/10.1049/iet-its.2018.0064, eprint: https://onlinelibrary.wiley.com/doi/pdf/10.1049/iet-its.2018.0064
38. Ni, W. & Cassidy, M. Cordon control with spatially-varying metering rates: A Reinforcement Learning approach. *Transportation Research Part C: Emerging Technologies.* **98** pp. 358–369 (2019,1), https://www.sciencedirect.com/science/article/pii/S0968090X18312592
39. Zhou, D. & Gayah, V. Scalable multi-region perimeter metering control for urban networks: A multi-agent deep reinforcement learning approach. *Transportation Research Part C: Emerging Technologies.* **148** pp. 104033 (2023,3), https://www.sciencedirect.com/science/article/pii/S0968090X23000220
40. Korecki, M., Dailisan, D. & Helbing, D. How Well Do Reinforcement Learning Approaches Cope With Disruptions? The Case of Traffic Signal Control. *IEEE Access.* **11** pp. 36504–36515 (2023), https://ieeexplore.ieee.org/document/10100954, Conference Name: IEEE Access
41. Aslani, M., Seipel, S., Mesgari, M. & Wiering, M. Traffic signal optimization through discrete and continuous reinforcement learning with robustness analysis in downtown Tehran. *Advanced Engineering Informatics.* **38** pp. 639–655 (2018,10), https://www.sciencedirect.com/science/article/pii/S1474034617302598
42. Rodrigues, F. & Azevedo, C. Towards Robust Deep Reinforcement Learning for Traffic Signal Control: Demand Surges, Incidents and Sensor Failures. *2019 IEEE Intelligent Transportation Systems Conference (ITSC).* pp. 3559–3566 (2019,10), https://ieeexplore.ieee.org/document/8917451
43. Wu, C., Ma, Z. & Kim, I. Multi-Agent Reinforcement Learning for Traffic Signal Control: Algorithms and Robustness Analysis. *2020 IEEE 23rd International Conference On Intelligent Transportation Systems (ITSC).* pp. 1–7 (2020,9), https://ieeexplore.ieee.org/document/9294623
44. Lillicrap, T., Hunt, J., Pritzel, A., Heess, N., Erez, T., Tassa, Y., Silver, D. & Wierstra, D. Continuous control with deep reinforcement learning. (arXiv,2019,7), http://arxiv.org/abs/1509.02971, arXiv:1509.02971 [cs, stat]
45. Mnih, V., Kavukcuoglu, K., Silver, D., Rusu, A., Veness, J., Bellemare, M., Graves, A., Riedmiller, M., Fidjeland, A., Ostrovski, G. & Others Human-level control through deep reinforcement learning. *Nature.* **518**, 529–533 (2015)
46. Kim, H., Muñoz, S., Osuna, P. & Gershenson, C. Antifragility Predicts the Robustness and Evolvability of Biological Networks through Multi-Class Classification with a Convolutional Neural Network. *Entropy.* **22**, 986 (2020,9), https://www.mdpi.com/1099-4300/22/9/986, Number: 9 Publisher: Multidisciplinary Digital Publishing Institute
47. Zhang, R. & Zhang, J. Long-term pathways to deep decarbonization of the transport sector in the post-COVID world. *Transport Policy.* **110** pp. 28–36 (2021,9), https://www.sciencedirect.com/science/article/pii/S0967070X21001621

References

48. Bruijn, H., Größler, A. & Videira, N. Antifragility as a design criterion for modelling dynamic systems. *Systems Research And Behavioral Science*. **37**, 23–37 (2020), https://ideas.repec.org//a/bla/srbeha/v37y2020i1p23-37.html, Publisher: Wiley Blackwell
49. Axenie, C., Kurz, D. & Saveriano, M. Antifragile Control Systems: The Case of an Anti-Symmetric Network Model of the Tumor-Immune-Drug Interactions. *Symmetry*. **14**, 2034 (2022,10), https://www.mdpi.com/2073-8994/14/10/2034, Number: 10 Publisher: Multidisciplinary Digital Publishing Institute
50. Axenie, C. & Saveriano, M. Antifragile Control Systems: The case of mobile robot trajectory tracking in the presence of uncertainty. *IEEE Access*. **11** pp. 138188–138200 (2023), http://arxiv.org/abs/2302.05117, arXiv:2302.05117 [cs, eess]
51. Sun, L., Makridis, M., Genser, A., Axenie, C., Grossi, M. & Kouvelas, A. Antifragile Perimeter Control: Anticipating and Gaining from Disruptions with Reinforcement Learning. (arXiv,2024,2), http://arxiv.org/abs/2402.12665, arXiv:2402.12665 [cs, eess]

Chapter 4
Induced Antifragility

Abstract This chapter is dedicated to induced antifragility. Here, we discuss the benefits of input distribution irregularities based on emergent system dynamics in a feedback loop with a controller that drives the system towards a prescribed dynamics. We consider methods for detecting, analyzing, modelling, and controlling road traffic, robotics, and industrial control systems' antifragility.

4.1 Antifragile Road Traffic Control

Traffic on urban roads is a complex process that is constantly changing. Large fluctuations in traffic flow can be caused by random events that spread over time and space. Fortunately, there are periodic phenomena that explain traffic despite its complicated mechanics and unpredictability. In this section, we approach traffic control from the perspective of antifragility. We consider, on the one hand, the assessment of traffic fragility and, on the other hand, the design of antifragile control systems. We support our analysis and theoretical development with quantitative experiments that demonstrate the benefits of antifragile control on traffic quality metrics.

4.1.1 Introduction

Despite its serious impact on infrastructure and the social and economic aspects of city life, urban congestion remains resistant to simple solutions. As highlighted in the assessment of [1], traffic modelling, control and optimization remain one of the most challenging topics across disciplines, with only incremental research driving incremental progress. Current traffic control systems, such as SCOOT by [2], SCATS from [3], PRODYN from [4], OPAC from [5], RHODES from [6], or LISA from [7], use a simple model of traffic dynamics and feed it with sensor-based vehicle detection to optimize signal timing. Detection from multiple intersections is then combined into a central system to simulate traffic patterns in the vicinity. To minimize unnecessary

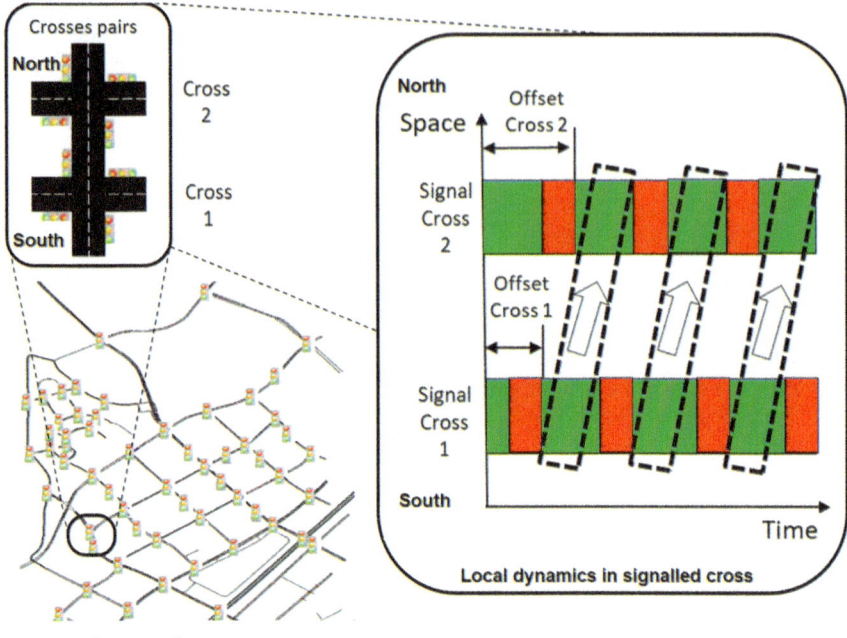

Fig. 4.1 Traffic control signaling. The space-time diagram of the vehicles' platoon traveling from South to North describes the basic dynamics in a pair of signalized crossings for the duration of the green light signal. The periodic signal behavior ensures that adjacent directions (i.e., West to East transit) are allowed to pass through (not shown in space-time). To provide for safe driving direction sequencing, the control algorithm should compute the green and red time signals while accounting for an offset

green phases and maximize efficient traffic flow, the underlying traffic model is then used to adjust the phasing of the traffic lights to match the traffic flow, as shown in Fig. 4.1.

4.1.1.1 Traffic Control

If all the temporal and geographical interactions between road junctions are understood and well-modeled, closed-loop road traffic control can theoretically be scaled up to the city level. But this is rarely the case, and other dimensions come into play. For example, there are weekly variations in traffic load patterns on fixed-capacity arteries (see work of [8]), Volatility of hourly capacity on multi-lane roads in city centers or during sports activities (see work by [9]), and variability of external traffic entering the city or heavy rainfall (see work of [10]) are clear components of traffic dynamics. These elements increase the likelihood of traffic congestion. The challenge is that such events are difficult to predict, model and manage. In technical terms, the

4.1 Antifragile Road Traffic Control

fundamental feature that distinguishes current systems is their traffic control model, or more specifically, the traffic dynamics they capture and how this model manages the inherent unpredictability, volatility and variability of the variables it captures.

4.1.1.2 Traffic Modelling

One technique that originates from physics that shows promise for traffic control is oscillator-based modelling and control. In a very good review and perspective, the study of [11] introduced the formalism of oscillator-based traffic modelling and control. The authors argue that such a method works well at the single cross-level, but it will not perform well in large-scale heterogeneous road networks due to factors including non-uniform road layout, interrupted traffic patterns, volatile traffic loads, and unpredictable weather. Our strategy in this work is to create a closed-loop antifragile control system using an antifragile controller designed for a non-linearly-coupled oscillators network model, based on the model of [12].

4.1.1.3 Fragility-Robustness-Antifragility Continuum in Traffic Control

The dynamics of traffic are highly nonlinear and sensitive to multiple sources of uncertainty. In this study, we go beyond uncertainty in capturing the real dynamics of traffic through a model (i.e., structured/parametric uncertainties) and consider the changes that disruptions, such as weather, accidents, social events, and infrastructure availability (i.e., unstructured uncertainties, or un-modelled dynamics), induce in the overall flow of cars. The inherent uncertainty, volatility, and variability of such disruptions are described by a stochastic evolution in space, time, and intensity. The compound effect of such disruptions (typically additive in nature) is reflected in computed measures of the quality of traffic, for instance, average speed, time loss, waiting time, or travel time for cars over a certain itinerary. Such unstructured uncertainties result in the re-computation of traffic signals, which subsequently alter the shape of the travel time distribution and, consequently, the overall travel time distribution—as depicted in Fig. 4.2.

The change in travel time can be described by how people respond to uncertainty. We can then describe how people respond to uncertainty-triggered signal re-computation concerning the travel time itself. As shown in Taleb's excellent work (2022), if we combine the uncertainty response with the distribution of the travel time, we can describe the probability distribution of the signal re-computation. We can change the signal re-computation parameters to change the shape of $S(TravelTime)$ to handle changes in $P(TravelTime)$, as shown in Fig. 4.3.

The degree of uncertainty affects how traffic flows. In traffic engineering, this is shown by the Macroscopic Fundamental Diagram (MFD). As systematically introduced in the work of [13], MFD describes relationships between traffic flow, density, and speed that can be used to map to the fragile-robust-antifragile spectrum. In our

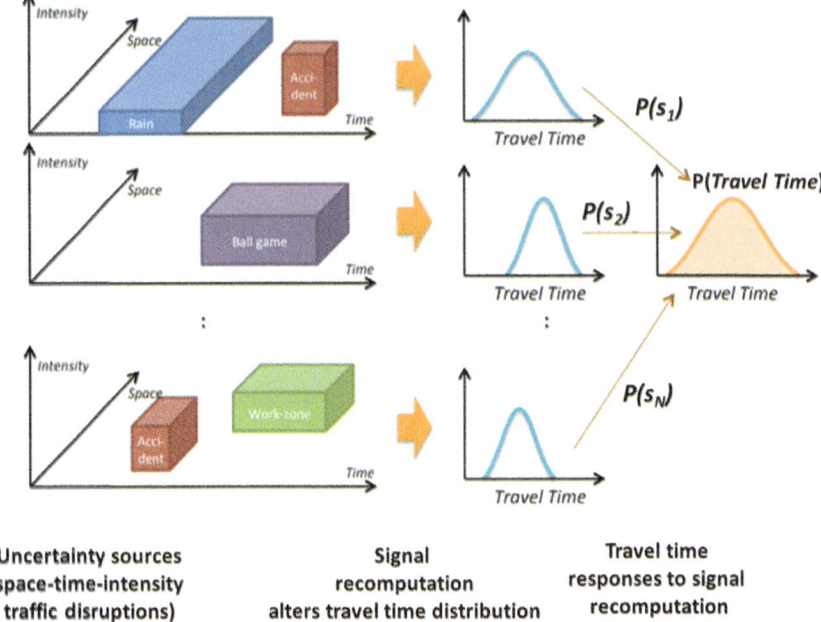

Fig. 4.2 Uncertainty, volatility and variability in traffic control when considering travel time over a journey. The space-time intensity of different sources of uncertainty in traffic (e.g. weather, accidents, social events, infrastructure availability) and the traffic signal recalculation that changes the travel time distribution. The change in the overall (whole itinerary) travel time distribution in response to the signal re-calculation

study, we propose a new alignment among the two spectra, depicted in Fig. 4.4. We consider the theoretical MFD, the three possible equilibrium states, and their placement in the velocity-density, flow-velocity, and flow-density characteristics (see Fig. 4.4a).

As one can see in Fig. 4.4b, the real MFD shows that both capacity drops and concave-to-convex MFD shapes are common in practice. The extracted shape from the data follows the analytic shape, so we can identify a direct mapping. The free flow region corresponds to a fat tail in the gain domain (increasing velocity with sub-linearly increasing flow) and a thin tail in the loss domain (high velocity and high flow). This is practically the congestion region of the MFD. The robust behavior is seen in the region where moderate velocity gives the most cars (i.e., bound flow). This is characterized by a thin tail in the loss domain and a thin tail in the gain domain. The real data match the analytic shape for velocity and density. The velocity decreases towards a creeping regime where congestion forms. The distribution of the characteristic's antifragile region shows the high-velocity average density of the MFD. Uncertainty can only push the system across the robust region. At the other end of the spectrum, we have the fragile region where congestion forms with a fat

4.1 Antifragile Road Traffic Control

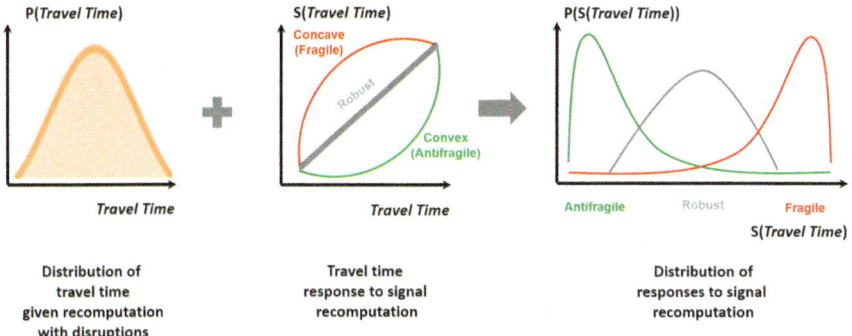

Fig. 4.3 Mapping from uncertainty effects on travel time $P(TravelTime)$ to the fragile-robust-antifragile spectrum under signal re-computation. Signal re-computation response of travel time changes shape $S(TravelTime)$ can vary based on the closed-loop control variables and push the system towards a different region of the response to signal re-computation distribution $P(S(TravelTime))$. Due to the periodic nature of the traffic light signaling the shape of $S(TravelTime)$ can be convex (i.e., antifragile behavior), concave (i.e., fragile behavior), linear (i.e., robust behavior), or even mixed convex–concave (i.e., non-stationary behavior)

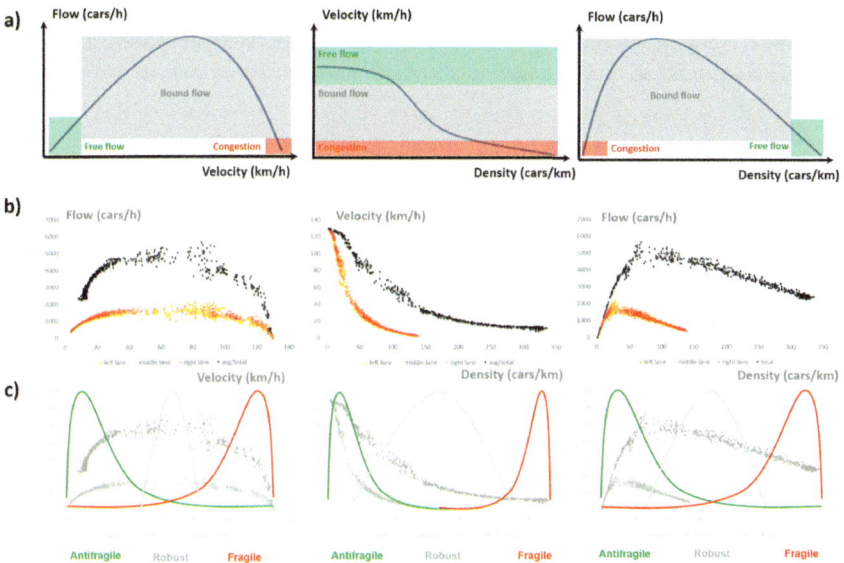

Fig. 4.4 Mapping macroscopic fundamental diagrams (MFD) to the fragile–robust–antifragile spectrum in road traffic control under uncertainty. **a** Analytic form of the MFD and the traffic regime characteristics. **b** Real MFD extracted from a real-world dataset on a highway segment with three lanes. **c** The fragile–robust–antifragile distribution shape based on the mapping of MFD on the real data curves. The shapes of the distributions in the loss and gain regions are matched using Taleb's heuristic (see [14]) to the fragile–robust–antifragile continuum

tail. The flow-density characteristic shows that traffic dynamics without control has limits. Figure 4.4b shows that the flow of cars doesn't decrease at high densities. This means the system reaches the fat tail in the loss region.

4.1.2 Induced Traffic Antifragility

The example shows why a control algorithm is needed to push closed-loop system trajectories to the antifragile regions under the effect of the traffic signal timing. We extend the fragility-robustness-antifragility detection with degrees of fragility (inherited and intrinsic fragility as defined in [14]). To realize induced antifragility, a closed-loop control system must be designed to compensate for uncertainty.

As we see in our simple analysis in Fig. 4.4, intrinsic and inherent fragility characterize the open-loop traffic dynamics, where the second-order effects (i.e., the free flow, bound flow, and congestion) of the MFD can be mapped to the fragile–robust–antifragile spectrum Fig. 4.4c. Without a traffic light, the system's response to uncertainty is based on the shape of the probability distribution of its variables and how sensitive it is to uncertainty. The characteristics of uncertainty also define inherited fragility and consider the combined dynamics of the system and uncertainty.

Our study looks at how to make traffic flow more smoothly, which is depicted in Fig. 4.5.

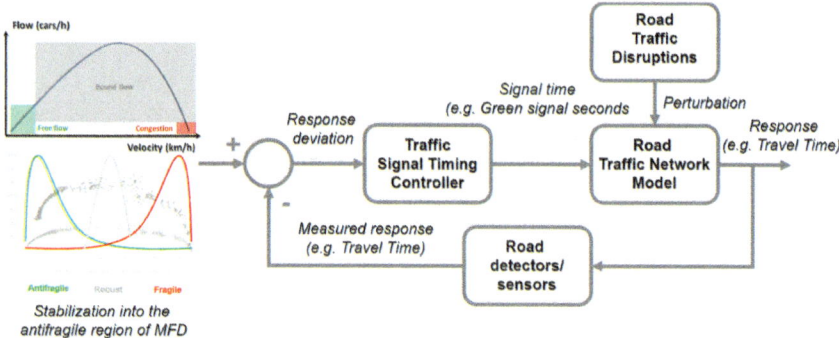

Fig. 4.5 Antifragile control closed–loop system. The idea is to make the road traffic network model more stable in the antifragile region of the MFD. This means that if there are disruptions to traffic, the control law (i.e., signal time) can make the system respond (e.g. travel time) in a way that is within the antifragile region of the MFD

4.1.2.1 System Design Methodology

This section presents the models and tools employed in the study. It begins with an oscillator-based network model of a large urban road traffic scenario. Subsequently, the formal introduction of the antifragile control framework for the road traffic scenario is provided. Finally, we introduce the formalism and parametrization of alternative (and relevant) road traffic control and optimization methods (i.e., Optimal Control (using Mixed Integer Linear Programming (MILP)) and Robust Control from [12]) to comparatively evaluate and probe the benefits of antifragile control. Both road traffic and road traffic control signals exhibit strong periodic behavior. On fast timescales, the periodicity of traffic light signals determines fluctuations in the road traffic flow, which in turn manifest as daily periodic patterns (i.e., slow timescales). One may, for instance, observe the "camel-back" profile of traffic over working weekdays, which peaks around 9 AM and 6 PM. This observation motivates us to describe the traffic light phasing phenomenon as a repeated synchronization problem, in which a large network of oscillators of [12] is used to model traffic interactions. The underlying rationale is as follows: (1) Each of the traffic light oscillators is injected with traffic flow data, as measured by sensors, which impacts its local dynamics. This is to say that the local computation of green time is given by neighboring oscillators. (2) The oscillator network, which is mapped on the traffic lights of an urban region, converges to a steady state. This steady state is used to extract an adaptive factor, which is then used to adjust the traffic light phases. Despite the inevitable differences in the natural oscillation frequencies and the injected traffic flow data of each oscillator, the network ensures that each of the coupled oscillators repeatedly locks phase (i.e., synchronizes given the interactions with the neighboring oscillators within a cross and between crosses), as described in Eq. (4.1).

$$\frac{d\theta_i(t)}{dt} = \omega_i(t) + k_i(t) \sum_{j=1}^{N} A_{ij} sin(\Delta\theta) + F_i sin(\Delta\theta^*) \tag{4.1}$$

where
θ_i—the amount of green time of traffic light i
ω_i—the frequency of traffic light i oscillator
k_i—the flow of cars passing through the direction controlled by oscillator i
A_{ij}—the static spatial adjacency coupling between oscillator i and oscillator j
F_i—the coupling of external perturbations (e.g. maximum cycle time per phase)
θ^*—the external perturbation (e.g. traffic signal limits imposed by law) and
$\Delta\theta = \theta_j(t) - \theta_i(t)$
$\Delta\theta^* = \theta^*(t) - \theta_i(t)$.

The model posits that the change in allocated green time, represented by the variable θ_i, for a specific traffic light, i, in a given direction, depends on the following factors: The model considers three factors: (1) the internal frequency of the corresponding traffic light oscillator, represented by the variable ω; (2) the current flow of cars, represented by the variable k_i; and (3) the spatial coupling, represented by the variable A_{ij}, which describes the impact of a nonlinear periodic coupling of the oscillators through the street network. The oscillators are also influenced by the following factors: (1) the phase difference between the oscillators, represented by the function $sin(\theta_j(t) - \theta_i(t))$; (2) the external perturbation, represented by the function with weight F_i, which ensures that the output of the system remains within the bounds of realistic green time values as defined by traffic laws. In light of the established topological configuration of the road network and the calculated green times for each oscillator, it is possible to infer the actual adaptive factor (i.e., delta green time) to be applied to the traffic light phase duration between adjacent oscillators corresponding to adjacent moving directions. This is based on the convergence of the dynamics (i.e., the solution of the differential equation referenced as 4.1) and the steady state value of the green time (i.e., the solution $\theta_i(t_f)$), we calculate the phase duration as the time to synchronization of each oscillator relative to the ones coupled to it. From the dynamics synchronization matrix ρ at each time t the phase duration update is calculated as $\arg\max_t\{\rho(t) > \tau\}$ where $\rho_{ij}(t) = cos(\theta_i(t) - \theta_j(t))$ and $0 < \tau < 1$.

To ground the analytic formulation, we describe a simple, regular 5×5 lattice composed of $N = 25$ oscillators which describe an ideal road network. For the sake of simplicity, in this example, each oscillator is responsible for an entire cross (i.e., the four adjacent directions: N, S, W, E) and the spatial coupling A_{ij} is given by the topology of the lattice, as illustrated in Fig. 4.6. In this case, the dynamics of each oscillator, designated by the index i, is represented by the superposition of its intrinsic oscillation frequency, represented by the variable ω_i, and the collective influence of neighboring oscillators, coupled through the matrix A_{ij}, weighted by the flow of vehicles, represented by the variable k_i through the cross controlled by oscillator i. Figure 4.6b describes the internal dynamics of such a network model for traffic control where given the different initial conditions of each oscillator, the coupling dynamics enforces consensus after some time (i.e., $\approx 2.1s$). The steady state is then used to extract the actual phase duration by simply calculating the time to synchronization $\arg\max_t\{\rho(t) > \tau\}$, as a per oscillator relative time difference, from the ρ matrix in Fig. 4.6c. Here, the choice of τ determines how fast a suitable steady state is reached.

4.1.3 Antifragile Traffic Control

The objective of our control-theoretic approach is to induce a shift in a system's behavior from a fragile or robust region of the spectrum to the antifragile region of the MFD. It is acknowledged that the system is inherently fragile and that this is an immutable characteristic. The objective is to induce antifragility through the application of a control law that guides the system trajectories towards the desired region. The primary tenets of this approach, and the fundamental principles of induced antifragility, are as follows: (1) *redundant overcompensation*, whereby the system

4.1 Antifragile Road Traffic Control

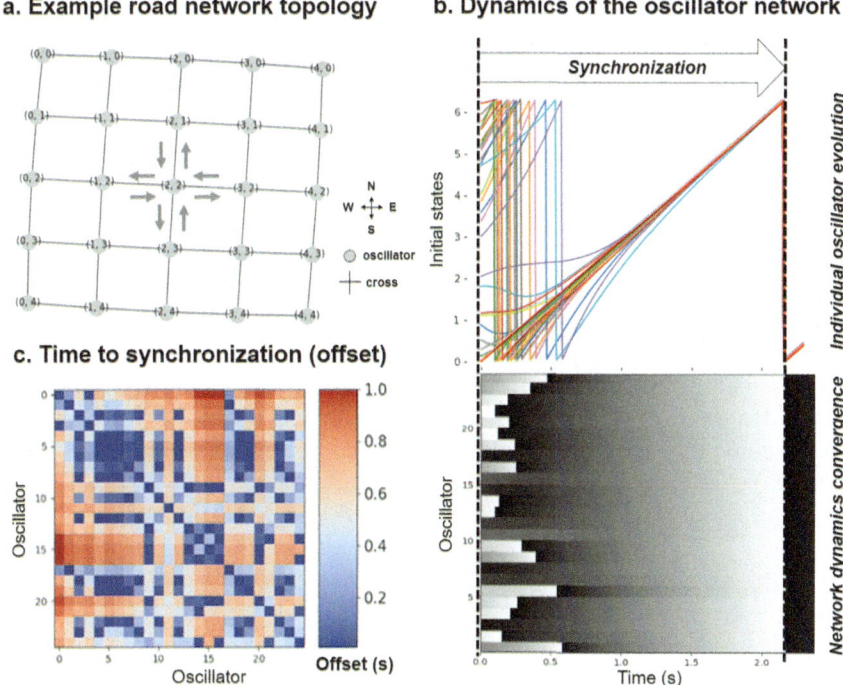

Fig. 4.6 Oscillator-based network model dynamics. **a** Sample network geometry and spatial coupling of oscillators. **b** Dynamics of the oscillator network model from initial conditions and without control signals, phase lock under "free evolution" after a transient. **c** Time to synchronization of one oscillator w.r.t. spatially connected neighbors as a quantification of offset of green time to compute the green time phase

trajectory overshoots in a controlled manner to build extra capacity in anticipation; and (2) *structure-variability*, achieved through A discontinuous control law that can induce high-frequency stress while still driving the system's trajectory in the desired region; and (3) the *convexity of the system response* once it has reached the desired region under the effect of the control law.

4.1.3.1 Formalism of Induced Antifragile Control Design

In a previous study [15], which established the foundations of the control-theoretic approach to induced antifragility, we employed the geometric control and Riemannian geometry objects, as formally described in [16]. This enabled us to work in a coordinate-free space, relying on the embedding of a manifold into a larger dynamical space that allowed for simpler control law definitions suitable for manifolds with curvature (i.e., second-order effects). In the current study, we consolidate this framework and relax some of the assumptions previously made.

Preliminaries

In our mathematical development of the study, we will always consider the most general form of the (oscillator-based) traffic model as the dynamical system in Eq. (4.1), namely

$$\dot{\theta}_i = f_i(t, \theta_i, u_i) \quad (4.2)$$

where the state vector θ_i takes values on a smooth manifold Θ, t is the time, u_i is the control law for oscillator i, and f_i is a smooth nonlinear function. In our initial realization of antifragile control in [15], we framed our problem as a tracking control problem where a reference (i.e., desired) vector θ_{id} (i.e., the desired dynamics $\theta_i = \theta_{id}$) was used to compute an error function $\varphi(\theta_i, \theta_{id})$ subsequently used in the variable structure control design, as described in [17, 18]. More precisely, the design objective was the synthesis of a control law u such that the system trajectories reach and stay on a manifold (i.e., the desired region on the MFD in the fragile–robust–antifragile spectrum) σ, in other words, $\sigma(t, \theta_i, \theta_{id}) = 0$, for which

$$\varphi(\theta_i, \theta_{id}) = \theta_i - \theta_{id} \quad (4.3)$$

$$\sigma(\theta_i, t) = \left(\frac{d}{dt} + \lambda\right) \varphi(\theta_i, \theta_{id}), \lambda > 0 \quad (4.4)$$

This allowed us to deal with uncertainty by reaching and staying under the effect of u_i on σ, which now describes the overall system dynamics. In other words, given that $\theta_{id}(0) = \theta_i(0)$, our control problem of tracking $\theta_i = \theta_{id}$ is equivalent to that of remaining on $\sigma(t, \theta_i)$ for all $t > 0$. Opposite to our initial realization of antifragile control, we replace the tracking problem $\theta_i = \theta_{id}$, or in other words $\lim_{t\to\infty} \varphi(\theta_i, \theta_{id}) = 0$, with a stabilization problem in σ. We compute the manifold σ as a discontinuous element (i.e., variable structure assumption still holds) and its second derivative $\ddot{\sigma}$ as a function of the rate of change of the control law u. This allows us to drop the dependency on $\varphi(\theta_i, \theta_{id})$ and many other benefits, as we will see in the following section. Secondly, the redundant overcompensation feature of antifragile control is now implemented on a generalized basis through the use of a variable structure (attractor dynamics) controller, as opposed to a manifold proportional derivative (PD) control and a problem-dependent The selection of a Riemannian transport map (i.e., a description of the actual dynamics transformation that guides the system towards the desired manifold along a geodesic, in this case, the fragile-robust-antifragile spectrum of the MFD) is, therefore, a key consideration. Using attractor dynamics allows us to reduce the chattering effects (i.e., excessive switching while approaching the σ) by using well-behaved dynamics given by the second-order differentiation of the attractor manifold $\ddot{\sigma}$ that depends on the rate of change of the synthesized control law $\dot{u}$. This depends proportionally on the curvature of the attractor manifold landscape as well. The aforementioned generalizations of the preceding iteration of antifragile control are predicated on two fundamental considerations. Firstly, the specific system under consideration, namely a network of coupled oscillators tasked with optimizing road traffic flow and minimizing travel

4.1 Antifragile Road Traffic Control

time, necessitates a particular approach. Secondly, the inherent flexibility and precision afforded by second-order attractor dynamics offers a promising avenue for nonlinear control, as elucidated by the seminal work of [19, 20], and of [21]).

Problem formulation

To facilitate the desired motion of the closed-loop system, we consider a stabilization problem about the system's dynamics, represented by the function $\dot{\theta}_i$, on a known manifold (i.e., attractor surface) σ. However, to guarantee the precision of the motion (i.e., attractor) and to circumvent the implementation of high-frequency control actions that could prove detrimental to traffic control, we propose the utilization of high-order motion manifolds (i.e., attractor surfaces) and the synthesis of an attractor controller following the recommendations outlined in [17, 18]. This type of control is an excellent candidate for achieving closed-loop antifragility, as evidenced by the preliminary work of [15]. We will present the theoretical foundations of the control synthesis in the following section. For the sake of simplicity, we will replace θ_i with θ throughout this section, while maintaining the essence of the single oscillator dynamics within the network.

As previously stated, the tracking problem of the desired dynamics, represented by the variable θ_{id}, is converted into a stabilization problem in σ. This is readily apparent upon differentiating σ once in Eq. (4.4) for the control law u_i to emerge (that is, simply differentiate Eq. (4.4) and replace $\dot{\theta}_i$ with Eq. (4.2))). This is also an intuitive approach, as no individual green time is available, but only the total allocated green time within a cross/region. Stabilizing to σ accounts for driving the network of oscillators to a phase lock within a specified time frame, which can be defined as reaching time or synchronization time. Furthermore, bounds on the attractor manifold σ can be directly translated into bounds on the tracking error vector $\varphi(\theta_i, \theta_{id})$, and therefore σ represents a true measure of tracking performance, as demonstrated in [18].

The objective is to ascertain the high-order effects of the attractor surface σ, to determine both the motion of the closed-loop system (i.e., phase lock on an optimal configuration of traffic light green time allocations) and the control law signal u (i.e., total green time reaches police-imposed values). Let's consider the l-th derivative of σ as being

$$\sigma^{(l)}(\theta, t) = \left(\frac{d}{dt}\right)^{(l)} \sigma(t, \theta(t, \epsilon)) \tag{4.5}$$

a uniformly bounded function of ϵ. Then for the steady state part of $\theta(t, \epsilon)$ there exist positive constants $C_1, C_2, ..., C_{l-1}$ such that the following equations hold

$$\|\dot{\sigma}\| \leq C_1 \tau^{l-1} \tag{4.6}$$

$$\|\ddot{\sigma}\| \leq C_2 \tau^{l-2} \tag{4.7}$$

$$\vdots \tag{4.8}$$

$$\|\sigma^{(l-1)}\| \leq C_{l-1} \tau \tag{4.9}$$

if $\tau(\epsilon) > 0$ is the smallest time interval of smoothness of the piece-wise smooth function $\theta(t, \epsilon)$. This demonstrates that considering high-order attractor manifolds allows for a smooth control signal. This will turn out useful in our traffic control problem where the choice of green time needs to judiciously follow a smooth pattern to allow connected oscillators to lock phase. If we consider Eq. (4.2) in Filippov's sense (see work of [22] for formalism), we can then replace it by an equivalent differential inclusion $\dot{\theta} \in F(\theta, t, u)$ where if the vector field F is continuous then the set value $F(\theta_0, t_0, u_0)$ is the convex closure, as demonstrated in the work of [23]. We now integrate this new formulation in the framework initially introduced in [15] where the Riemannian geometry offers the mathematical framework to develop the control synthesis.

Let Γ be a smooth manifold. The set Γ is called a first-order point attractor. The second-order attractor point set is the set of points $\theta \in \Gamma$, where $F(\theta)$ lies in the tangential space $T_\theta \Gamma$ to the manifold Γ at the point θ. We remind the reader that T_θ is the parallel transport map that describes the actual dynamics transformation pushing the system towards the desired manifold along a geodesic. Then there exists a second-order attractor on Γ in the vicinity of a second-order point attractor θ_0 if, in this vicinity of the point, θ_0 is an integral set (i.e., it consists of solutions in the Filippov sense—see [23]).

In this case, all possible velocities F lie in the tangential space $T_\theta(F)$, and even when a switching error is present, the state trajectory is tangential to the manifold Γ at the time of leaving. Back to the formulation in Eq. (4.2), we can now extend the description of the closed-loop as following

$$\dot{\theta} = f(t, \theta, u) \tag{4.10}$$

$$u = U(t, \theta, u) \tag{4.11}$$

$$\xi = \Psi(t, \theta, \xi) \tag{4.12}$$

where U now describes the first-order attractor/line attractor (i.e., recall that differentiating σ one recovers u) and ξ describes the second-order attractor, basically the velocity of the control law u. Importantly, in second-order attractor dynamics U is a continuous function and Ψ is bounded discontinuous, hence the formulation can be written as a continuous control

$$\dot{\theta} = f(t, \theta, u_{eq}(t, \theta)), \tag{4.13}$$

where u_{eq} is the equivalent control law evaluated from

$$\dot{\sigma}(t, \theta) = \frac{\partial}{\partial t}(\sigma(t, \theta)) + \frac{\partial}{\partial \theta}(\sigma(t, \theta))f(t, \theta, u_{eq}) = 0 \tag{4.14}$$

4.1 Antifragile Road Traffic Control

or

$$\dot{\sigma}(t, \theta, u) = L_u \sigma(t, \theta) \tag{4.15}$$

where

$$L_u(\cdot) = \frac{\partial}{\partial t}(\cdot) + \frac{\partial}{\partial \theta}(\cdot) f(t, \theta, u) \tag{4.16}$$

is the total Lie derivative with respect to $\dot{\theta} = f(t, \theta, u)$ when u is constant. Then, we can see that the curvature of the line attractor/manifold σ can be computed by

$$\frac{d^2}{dt^2}\sigma = \frac{d}{dt}(L_u \sigma) = L_u L_u \sigma + \frac{\partial}{\partial u} L_u \sigma \dot{u} \tag{4.17}$$

If we let $C = L_u L_u \sigma$ and $K = \frac{\partial}{\partial u} L_u \sigma$, then we can write Eq. (4.17) as

$$\ddot{\sigma} = \frac{d^2}{dt^2}\sigma = C + K\dot{u} \tag{4.18}$$

with $\|C\| \leq C_0$ and $0 < K_m \leq K \leq K_M$.

The solution that keeps the system on the attractor manifold in motion is obtained from $\ddot{u} = 0$ from Eq. (4.9). Hence, the velocity of the control law $\dot{u}$ is given by

$$\dot{u} = \begin{cases} -u & \text{if } |u| > 1, \\ -\alpha \operatorname{sign}(\dot{\sigma} - g(\sigma)) & \text{if } |u| \leq 1, \end{cases} \tag{4.19}$$

where $g = -\beta \operatorname{sign}(\sigma |\sigma|^\gamma), 0.5 \leq \gamma \leq 1$. The positive sub-unitary parameters α and β represent the *overcompensation factor* and the *anticipation factor* respectively. Additionally, we can have finer control when the system moves close to the attractor manifold σ by considering a boundary Φ so that the manifold reaching (i.e., attractor dynamics) is higher for $\dot{\Phi} < 0$ and lower for $\dot{\Phi} > 0$. Then we can rewrite the control law velocity $\dot{u}$ by replacing the sign function with a *sat* function such that $\operatorname{sign}(\sigma)$ can be replaced by $sat(\frac{\sigma}{\Phi})$ and then

$$\dot{u} = \begin{cases} -u & \text{if } |u| > 1, \\ -\alpha sat\left(\frac{\dot{\sigma}}{\Phi} - \beta \frac{\sigma |\sigma|^\gamma}{\Phi^2}\right) & \text{if } |u| \leq 1. \end{cases} \tag{4.20}$$

These results are consistent with the seminal work and study of [19] and the work of [24] in that the approach guarantees the finite-time convergence to a predefined real attractor manifold (i.e., a line with state-space attractors aligned on a first-order curve) with controllable overcompensation through the curvature (i.e., second derivative) of the attractor manifold. Line attractor curvature depends on the velocity of the control law $\dot{u}$, and implicitly determines the reaching time under u. In the following, we motivate the need for structure variability and attractor dynamics to reach closed-loop antifragile behavior. We now particularize the theoretical apparatus for the network of oscillators in Eq. (4.1).

Structure Variability

Recall that our goal is to design a line attractor to which the controlled closed-loop system trajectories must belong. Variable structure control, and more precisely attractor dynamics, is our candidate for the practical realization of antifragile control. Let's now consider the properties of line attractors in some greater detail.

First, in order to detect antifragility through the use of attractor dynamics control, the trajectories of the system's state vector must belong to manifolds of lower dimension than that of the whole state space, therefore the order of differential equations describing the motion on the attractor is also reduced. We previously demonstrated that 1-st order attractor control can drive closed-loop nonlinear systems to the antifragile region of their dynamics using a discontinuous control law (see [15] for details) using redundant overcompensation and a high-frequency control law. In the current work, we extend the idea to 2-nd order attractor dynamics and the impact that the curvature of the line attractor has upon closed-loop system motion. Most of the work in this area is covered in the excellent review of [21]. We start by considering the basic 2-nd order variable structure control synthesis where the α and β, namely the *overcompensation factor* and the *anticipation factor* are updated to confine the closed-loop system to the higher derivative of the control variable.

$$\ddot{\sigma} = C + K\dot{u}, \quad u(t) = -\alpha(t)U sign(\sigma - \beta\sigma_M) \quad (4.21)$$

where $\beta \in [0, 1)$ and,

$$\alpha(t) = \begin{cases} 1 & \text{if } (\sigma - \beta\sigma_M)\sigma_M \geq 0 \\ \alpha^* & \text{if } (\sigma - \beta\sigma_M)\sigma_M < 0, \end{cases} \quad (4.22)$$

with $\alpha^* \in [0, \frac{\beta}{10})$ and σ_M is the last extremal value of σ, typically $\sigma_M = \sigma(t_i)$, where t_i is the initial time instant.

Secondly, in the majority of practical systems, the motion on the attractor is independent of control and is determined solely by the properties of the model and the position (or equations) of the discontinuity surfaces. This allows the closed-loop system undergoing motion to achieve antifragile behavior. Moreover, the initial problem can be decomposed into distinct, lower-dimensional sub-problems, where the control is simply allocated to constructing an attractor. The requisite character of motion across the intersection of discontinuity manifolds is determined by an acceptable choice of their equations. In the context of antifragile control synthesis, the consideration of higher-order attractor dynamics allows for the attainment of more smoothly evolving dynamics in the vicinity of the line attractor. This assumption is based on the premise that, when the order is increased, an additional constraint is typically introduced, which is a linear combination of the original constraint (i.e., line attractor) and its successive total time derivatives. This concept has been demonstrated in

4.1 Antifragile Road Traffic Control

the work of [20]. In this formulation, the successive derivatives of σ do not depend on the control u, and $\sigma^{(l)}$ depends linearly on u or its velocity with non-zero coefficients $L_a^{(l-1)} L_b \sigma$ given that the system dynamics in Eq. (4.2) is rewritten explicitly as $\dot{\theta} = a(\theta) + b(\theta)u$ and L_a, L_b are the total Lie derivatives with respect to a, b, the state matrix, and the input matrix, respectively. This brings us close to a very important result, namely

$$\dot{\theta}(t) = a(t, \theta) + b(t, \theta)u(t), \sigma = \sigma(t, \theta) \tag{4.23}$$

$$\sigma^{(l)}(t) = h(t, \theta) + g(t, \theta)u(t) \tag{4.24}$$

where $h(t, \theta) = \sigma^{(l)}|_{u=0}$ and $g(t, \theta) = \frac{\partial}{\partial u} \sigma^{(l)} = 0$ supposing that for $K_m, K_M, C > 0$

$$0 < K_m \leq \frac{\partial}{\partial u} \sigma^{(l)} \leq K_M \tag{4.25}$$

$$\|\sigma^{(l)}|_{u=0}\| \leq C \tag{4.26}$$

is always true (at least) locally. The control law u would satisfy $u = \Psi(\sigma, \dot{\sigma}, \ddot{\sigma}, ..., \sigma^{(l)})$ hence for $\sigma = 0$ then $\Psi(0, 0, 0, ..., 0) = -\frac{h(t,\theta)}{g(t,\theta)}$. In other words we can write $\sigma^{(l)} \in [-C, C] + [K_m, K_M]u$ as a convex, semi-continuous function. This ensures that the closed-loop system trajectories in the plane $(\dot{\sigma}, \sigma)$ are confined between the trajectories of

$$\ddot{\sigma} = \pm C + K_m \Psi(\dot{\sigma}, \sigma) \tag{4.27}$$

$$\ddot{\sigma} = \pm C + K_M \Psi(\dot{\sigma}, \sigma) \tag{4.28}$$

which provides the (homogeneous) control law u as

$$u = -\alpha sat \left(\frac{\dot{\sigma}}{\Phi} - \beta \frac{\sigma \sqrt{\sigma}}{\Phi^2} \right) \tag{4.29}$$

with $\alpha K_m - C > \frac{\beta^2}{2}$. This control law comprises two components: firstly, a piecewise discontinuous component, given by the expression $\beta \frac{\sigma \sqrt{\sigma}}{\Phi^2}$, and secondly, an energy dissipation component, given by the expression $\frac{\dot{\sigma}}{\Phi}$. The former determines the finite convergence time to the attractor manifold, while the latter ensures the stability of the system. This combination enables asymptotic convergence in the presence of "unmatched" perturbations and uncertainty, as demonstrated analytically in the work of [25]. Furthermore, this enables the closed-loop system to manage unstructured uncertainty (i.e., frequency-domain disturbances typical for oscillator-based dynamics) beyond the conventional robust formulation utilizing $H\infty$ weighting, as outlined in Ref. [26]. The analysis presented in Ref. [27], indicates that the control law described by Eq. (4.29), is capable of handling parametric uncertainty at low frequencies and unstructured uncertainty at high frequencies.

Finally, an additional distinguishing feature of attractor dynamics is that it can become indifferent to alterations in the dynamic characteristics of the controlled system in specific circumstances (e.g. traffic congestion). It is crucial to highlight that, in contrast to continuous systems with non-measurable disturbances where the invariance constraints necessitate the use of indefinitely high gains, the same outcome can be achieved in discontinuous systems through the utilization of finite control actions.

4.1.3.2 Control Synthesis

The network dynamics of the antifragile control (Eq. (4.1)) have been meticulously parameterized to accommodate the typical daily traffic profile. This is evident in Fig. 4.7, which illustrates that the model is capable of recuperating the lost time through a single cross to an acceptable value., around $70s$ (see Fig. 4.7b). In the case of traffic disruptions (e.g. accidents, sports events, or adverse weather conditions), the system cannot capture the fast-changing dynamics (i.e., steep derivatives) of the traffic flow (see Fig. 4.7a) and, hence, performs poorly, for instance in preserving an acceptable time loss (i.e., difference in the duration of a trip in the traffic-free vs. heavy traffic) over rush-hour (see Fig. 4.7b around 18:00). The example in Fig. 4.16 illustrates a limitation of such dynamic networked models, namely handling uncertainty. Be it structured uncertainty (e.g. sub-optimal choice of the internal oscillator frequency ω or a sudden time-varying topological coupling A_{ij} through trajectory re-routing) or unstructured uncertainty (e.g. un-modelled dynamics through the single use of $\dot{\theta}(t)$ and neglecting rate of change given by the Laplace operator $\ddot{\theta}(t)$), the system in Eq. (4.1) is unable to converge to a satisfactory solution given input k and coupling constraints.

To address this challenge, we extend Eq. (4.1) with an antifragile control law. We choose to systematically maintain the stability of the oscillatory dynamics by using a control approach that ensures consistent performance in the face of uncertainties. We use the result we obtained in Eq. (4.29) and particularize for the main state change in Eq. (4.1).

Given that the control law u_i is given by Eq. (4.29) and by rewriting the system as $\dot{\theta}_i = a_i(\theta_i) + b_i(\theta_i)u_i$, we then have

$$u_i = -\alpha sat\left(\frac{\dot{\sigma}}{\Phi} - \beta\frac{\sigma\sqrt{\sigma}}{\Phi^2}\right), \qquad (4.30)$$

$$a_i = \omega_i(t) + k_i(t)\sum_{j=1}^{N} A_{ij}sin(\Delta\theta) + F_i sin(\Delta\theta^*) \qquad (4.31)$$

with the control gain b_i given by time to synchronization of a certain oscillator (i.e., time offset).

4.1 Antifragile Road Traffic Control

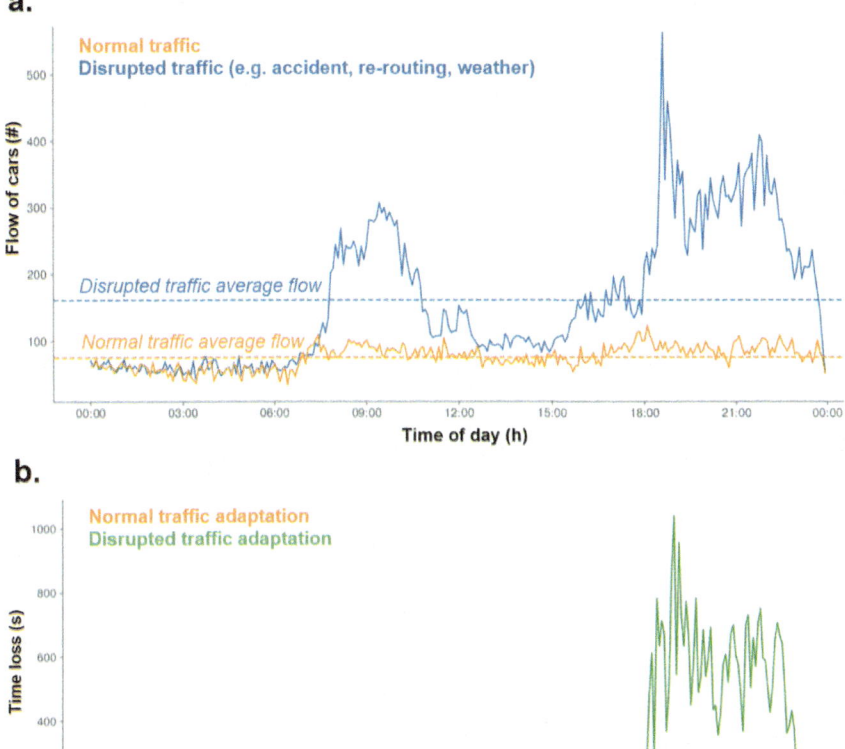

Fig. 4.7 Traffic dynamics in a daily urban scenario controlled by the model in Eq. (4.1). **a** The traffic flow profile over a day in a Chinese town (8 intersections) when considering the network of oscillators model. **b** Time loss during normal traffic over a day in a Chinese town (8 intersections) when considering the network of oscillators model. Changes in scale propagate in time when disruptions are present

This choice, in essence, captures and controls the impact of higher-order motion (i.e., second derivative) through a high-frequency switching of the control law towards synchronization. Such a discontinuous antifragile control "drives", through a regularizing control law term $u_i(t)$, the coupled dynamics of the oscillators towards a desired dynamics (i.e., line attractor σ).

4.1.3.3 Interactions Among the Components of the Antifragile Controller

The objective of the control law u_i is to exert a force on the network of coupled oscillators with a step size of δ, thereby guiding the system towards a state of dynamic equilibrium that accommodates the disruptions in the flow of cars k (i.e., captured by $\ddot{\theta}_i(t)$). This assumption is based on the intuitive notion that the controller is capable of capturing the second-order motion (i.e., the acceleration of the oscillator (i.e., $\ddot{\theta}_i(t)$)) and compensating for it asymptotically until the surface σ is reached. This assumes, initially, the selection of an appropriate attractor manifold, denoted by σ, which minimizes the quantity $b_i(t)$, as illustrated in Fig. 4.16c. The regularizing control law, $u_i(t)$, applied to oscillator i is the area under the curve (i.e., the integral) of the estimated energy surplus, depicted in Fig. 4.17c illustrates that the estimated surplus energy, which acts to maintain oscillator i at a distance from the desired antifragile dynamics ($s_i(t)$), is dependent upon two factors: the local oscillators' interaction (summation of the difference between $s_j(t)$ and $s_i(t)$, namely $\sum_{i,j}(\hat{s}_j(t) - \hat{s}_i(t))$) and the actual surplus energy. The change in surplus energy represents the actual dynamics of convergence to the line attractor and is based on the cumulative impact of neighboring oscillators (summed over all j) and the Laplacian of the green time (weighted by the direction of the convergence, $\text{sign}(s_i(t))$). The property of insensitivity of the line attractor to oscillatory dynamics is employed to regulate the response of the network of coupled oscillators to uncertainty. This was achieved in practice by incorporating the regularizing term, $u_i(t)$, into the local dynamics of each oscillator, as described by Eq. 4.1.

To get a better understanding of Eq. (4.31), we now exemplify, in Fig. 4.8, the impact the attractor dynamics controller has upon the dynamics of a road network when facing traffic disruptions from a real scenario (details about the data are provided in the Experiments and Results section). We consider a region composed of 8 crosses and $N = 29$ oscillators as described in Fig. 4.8a. In our case, the network of coupled oscillators is a system with discontinuous control (i.e., the control law $u_i(t)$ uses the sign of the energy surplus to drive the system towards the antifragile dynamics). As shown in Fig. 4.8b, c, given each sample of flow data $k_i(t_i...t_j...t_k)$ (from the road sensors) there is a fast convergence time scale which allows the oscillators to reach a steady state. This state is reached under antifragile control by compensating for the disruptions in the traffic flow modelled by the second-order motion $\ddot{\theta}_i(t)$. The stationary state is subsequently probed for the actual phase duration, relative to each coupled oscillator by solving $\arg\max\{\rho(t) > \tau\}$ where $\rho_{ij}(t) = \cos(\theta_i(t) - \theta_j(t))$ and $0 < \tau < 1$. Due to the fast changes occurring during disruptions (see Fig. 4.8b—rush hour around 6 PM), in the slow time-scale of traffic flow (i.e., sensory data), the network of coupled oscillators benefits from the antifragile control law to compensate for the abrupt changes and to reach consensus, as shown in Fig. 4.8c—right panel. This consensus state describes the point when the system dynamics reached the line attractor, in other words when the magnitude of the surplus energy decayed at a finite rate over the finite time interval (i.e., fast timescale in Fig. 4.8c—left panel).

4.1 Antifragile Road Traffic Control

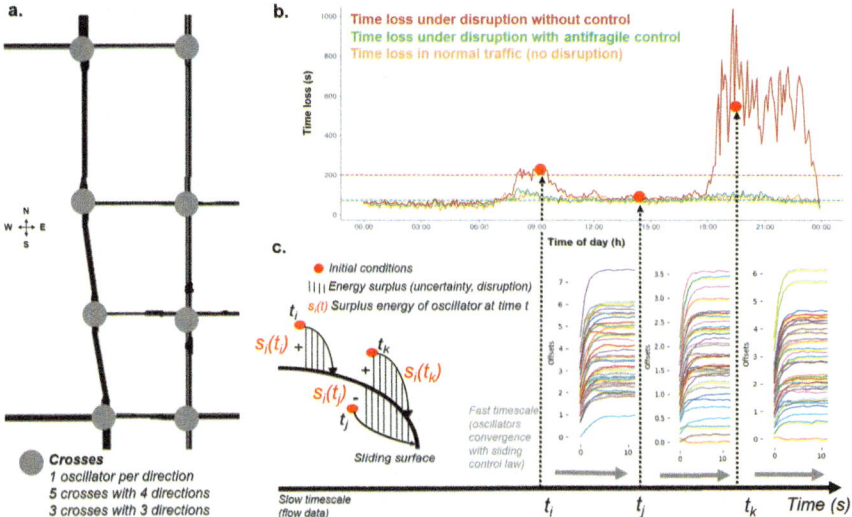

Fig. 4.8 The objective of this study is to present a novel approach to control for oscillator-based road traffic models, which we term "antifragile control". In each cross of the network under consideration, an oscillator (given by Eq. 4.1) controls the green time for each direction (i.e., N, W, S, E). A full day of traffic is used to evaluate the performance of the system with and without disruptions and with and without control. Each sample of measured traffic data is fed to the model, which converges rapidly to an estimate of green time given all spatial and temporal interactions with the other oscillators. **a** Road network geometry and associated oscillators. **b** Comparative time loss analysis of oscillator-based network model without disruptions, without control, and with antifragile control

Our proposed regularization approach has a simple physical interpretation. Uncertainty in the system behavior in the face of perturbations arises because the equations of motion of the dynamics in Eq. (4.1) are an ideal system model. Non-ideal factors such as un-modelled dynamics and sub-optimal parameter choices are neglected in the ideal model. However, by incorporating them into the system model, the ambiguity in the system behavior is removed and the system moves towards antifragile dynamics.

4.1.4 Experimental Results

The experiments and evaluation of the antifragile traffic controller utilize the SUMMER-MUSTARD (Summer season Multi-cross Urban Signalized Traffic Aggregated Region Dataset) real-world dataset, which comprises 59 days of authentic urban road traffic data from eight intersections in a city in China. The data employed in the experiments comprises traffic flows on each artery of the urban network, the spatial configuration of the network, and anonymized vehicle number plates tracked within the specified perimeter network (i.e., camera-based monitoring of vehicle movements for travel time). In particular, the data set includes information on traffic light cycle times, the number of vehicles in each cardinal direction (i.e., N, W, S, E) and the duration of green time at the traffic light. The road network layout underlying is depicted in Fig. 4.8a. To perform experiments and evaluate the system, we used the real-world traffic flows in the Simulator for Urban Mobility (SUMO) [28]. SUMO generates routes, vehicles, and traffic light signals that reproduce the real car flows in the dataset.

4.1.4.1 Antifragile Control Synthesis and Parametrization

It is noteworthy that the oscillator network is capable of accommodating spillback effects within a closed-loop feedback control system that is based on an antifragile controller. In particular, the proposed closed-loop control system displays characteristics that are analogous to those observed in Dynamic Traffic Assignment (DTA), which is currently a highly active area of research within the field of transport modelling [29]. In our design, we extend beyond the conventional static traffic assignment to encompass the formation and dispersion of vehicle queues at the modelling stage. This is incorporated into the Kuramoto model F_i term in Eq. 4.2. This is particularly the case in our scenario, where the optimization of travel times is a desired output along the traffic flow pattern. This is achieved by explicitly modelling queue length distributions in the F_i term of Eq. 4.2 and updating the overall model state to compute green time offsets. It can be demonstrated that as long as the term F_i in Eq. 4.2 provides an additive contribution to the periodic state, that is to say, $F_i sin(\Delta\theta)$, spillback will not disrupt the coupled oscillatory behavior that the model assumes. In our formulation, the Kuramoto model incorporates the finite spatial capacity of urban roads, which is of critical importance as this finite capacity exerts a profound influence on the evolution of traffic in a signalized network. Furthermore, the steady-state offset, which represents the network of coupled oscillators reaching a state of equilibrium, corresponds to the optimal solution within the spatiotemporal constraints of urban traffic, even in free-flow conditions.

4.1.4.2 Control Algorithms Evaluation

In the course of our experiments, we conducted a comparative evaluation of the adaptive behavior of the antifragile control and relevant state-of-the-art approaches against the static traffic planning (i.e., police-parametrized phases) that were used as the baseline. A comprehensive series of experiments was conducted, commencing with the real-world traffic flows recorded across the eight intersections in the SUMMER-MUSTARD dataset. To evaluate the adaptation and antifragility capabilities, a systematic introduction of progressive magnitude disruptions was implemented over the initial 59 days of traffic flow data. Such disruptions, including accidents and inclement weather, reduce velocity, which may lead to congestion. Furthermore, the occurrence of special activities, such as sporting events or the

4.1 Antifragile Road Traffic Control

commencement or conclusion of holidays, serves to augment the magnitude of the flow. The deterioration of traffic conditions may be attributed to a multitude of factors, including non-recurrent incidents such as accidents, adverse weather, or special events like football matches. By employing the real-world flow and SUMO, we have endeavored to reproduce the traffic flow behavior in the presence of disruption. Our approach involved utilizing normal traffic flow data, with the disruption effect on vehicle speed and network capacity and demand duly reflected. We have conducted a comprehensive sweep of disruption magnitude, encompassing a spectrum from normal traffic up to five levels of disruption, and have observed the effects across all eight crosses throughout the entire day.

The evaluated control systems are the following:

- BASELINE—An optimized static traffic planning that uses pre-stored timing plans computed offline using historic data in the real world.
- OPTIMAL—An optimal control method based on MILP phase plan optimization implementation inspired from [30].
- ROBUST—A basic implementation of a robust control based on a network of Kuramoto oscillators ([31]) for each direction in the road network cross from our previous work in [12].
- ANTIFRAGILE—The antifragile control with the attractor dynamics control law u.

The results of our evaluation are presented in Table 4.1, in which each approach is ranked on a scale of disruption magnitude (ranging from no disruption to maximum disruption) across a set of specific metrics (including average time loss, average speed, and waiting time). In the case of disruptions affecting flow magnitude, the level of disruption (i.e., 1.1–1.5) is used as a factor to adjust the number of vehicles or the speed of vehicles (i.e., in the event of adverse weather conditions) during the disruption. The evaluation was performed on the entire dataset, which comprised recorded traffic flows over 59 days from 8 locations.

The exhaustive experiments and evaluation in Table 4.1 demonstrate where the Antifragile Control Design method is an effective approach, although it does not consistently provide the optimal phase duration calculation. The selected evaluation metrics represent the comprehensive performance over an extended period, considering the most critical traffic metrics and the phase duration value estimated by each system.

The previous analysis is supported by the normalized ranking over the entire SUMMER-MUSTARD dataset in Fig. 4.9, where we provide a condensed visual representation of each system's performance. The average time loss metric reveals that the baseline approach exhibits the poorest performance, primarily due to its reliance on pre-defined timings and limited capacity to adapt to unanticipated disruptions within the daily traffic profile. At the opposite end of the ranking, both implementations of the ROBUST and the ANTIFRAGILE control provide minimal waiting time, effectively capturing rapid and pronounced changes in disrupted flows. As a consequence of their analogous core modelling and dynamics, the ANTIFRAGILE control systems and ROBUST exhibit comparable performance, with a relative

Table 4.1 A performance evaluation of the various phase duration calculation methods is presented herewith. The baseline method provides the nominal values, while the other methods record only the percentage improvement over the baseline

Control System/ Disruption level	Normal flow	1.1	1.2	1.3	1.4	1.5
Average time loss (s) BASELINE	102.53	214.60	236.22	341.38	385.39	393.11
Average time lost gain (%) OPTIMAL	25.31	28.34	26.01	34.23	43.017	42.46
ROBUST	35.09	24.57	30.30	30.98	39.797	37.12
ANTIFRAGILE	**47.67**	**48.84**	**39.72**	**44.16**	**39.88**	**44.29**
Average speed (km/h) BASELINE	45.81	35.67	30.46	27.02	24.94	19.75
Average speed gain(%) OPTIMAL	5.97	**7.57**	8.46	7.03	7.92	5.61
ROBUST	6.94	6.81	5.46	9.02	**8.29**	6.54
ANTIFRAGILE	**7.78**	5.97	**9.94**	**9.18**	5.07	**7.15**
Waiting time (%gain) BASELINE(s)	164.5	125.3	222.8	294.5	325.9	375.4
Waiting time gain(%) OPTIMAL	8.7	**22.7**	**21.8**	19.5	23.2	22.9
ROBUST	**15.7**	12.2	20.8	18.5	20.5	21.3
ANTIFRAGILE	10.7	145.7	20.8	**20.2**	**24.4**	**28.7**

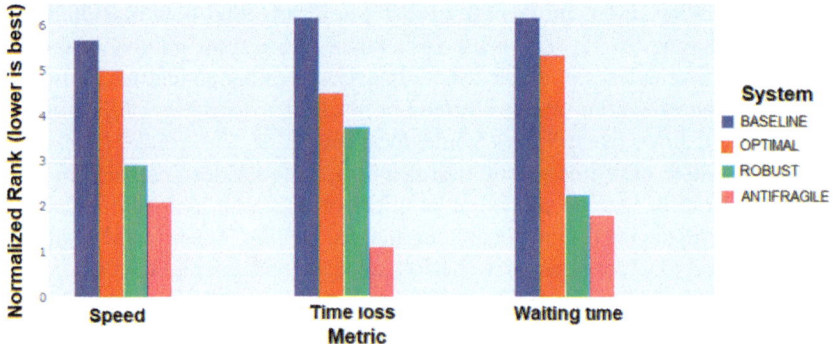

Fig. 4.9 Control systems ranking on all metrics and entire dataset (8 crosses over 59 days)

enhancement on the antifragile control side in terms of duration, waiting time, and speed metrics. This is due to its spatiotemporal extension beyond the basic oscillator model, which is capable of capturing the spatial contributions of adjacent flows beyond their temporal regularities when computing the phase duration. Finally, the ROBUST and the ANTIFRAGILE control system provide overall superior performance through its discontinuous attractor dynamics control law, which captures the deviation of the dynamics in the presence of disruptions and compensates robustly for their impact on the oscillator convergence (see Fig. 4.17).

4.1.4.3 Evaluation of the Run-Time

In terms of run time, the adaptive methods (i.e., OPTIMAL, ROBUST, and ANTIFRAGILE) provide different levels of performance, which can be attributed primarily to the modelling and optimization types employed. The baseline is excluded from consideration, as it represents merely the static, optimized plan allocation for the real traffic setting in SUMMER-MUSTARD. In essence, it is a rudimentary value recall from a look-up table. The time required by each of the evaluated adaptive systems to provide an estimate of phase duration following the acquisition of a sensory sample (i.e., a single reading of traffic flow data) was quantified. As previously stated, each system employs a distinct computational methodology. The optimal system utilizes a solver that implements an LP-based branch-and-bound algorithm, the robust system employs a Runge-Kutta 23 ODE solver, and the antifragile control system is implemented using a Runge-Kutta 45 ODE solver. The evaluation is presented in Table 2, which depicts the mean value across the full range of traffic conditions (normal and disrupted). The experimental setup employed three machines, each equipped with 24 CPU cores and 132 GB RAM, and Apache Flink for stream processing and cluster management (Table 4.2).

As anticipated, at the level of a single intersection, the OPTIMAL and ROBUST controls exhibit similar performance, resulting in a new phase duration value after 50 ms. At the regional level, when all eight crosses are considered, the run time increases by an order of magnitude. This is because the OPTIMAL control method overtakes the ROBUST control method, with the former displaying greater efficiency in optimizing constraints at scale and similar computations to the latter. The most expeditious approach, both at the single cross and regional levels, is antifragile control.

Table 4.2 Adaptive phase duration calculation run-time evaluation

MODEL	Single cross	Region (8 crosses)
OPTIMAL	0.0510	0.3930
Robust	0.0568	0.4544
ANTFRAGILE	**0.0071**	**0.0426**

With a run-time improvement exceeding 80% at both the single cross and regional levels, the antifragile control is distinguished by its efficient and straightforward computations.

4.1.5 Discussion

The conventional methodology for calculating green time for coordinated traffic signals is based on the average travel times between intersections and the average traffic volumes at each intersection. In practice, the "steep derivatives" of traffic flows (in open-loop/no-control) prevent the optimal and robust control from converging on the optimal phase duration value.

To achieve high performance, namely to minimize metrics such as time loss and to maximize average speed, we have synthesized an antifragile controller for the network of oscillators model. The controller in question exerts a force that propels the perturbed dynamics towards a state that drives the coupled oscillator's network towards an optimal phase. In this manner, the antifragile control system can address local and global traffic dynamics by leveraging the coupling among disparate oscillators that describe the traffic periodicity of disruptions.

As evidenced by our evaluation, OPTIMAL demonstrates efficacy in reducing route length; however, it is notable that the computational cost associated with the optimization process, which requires iterative convergence, is considerable. Ultimately, the ANTIFRAGILE control mechanism demonstrates superior performance in both speed and time loss metrics, which substantiates the assertion that a control law based on the second-order effects of signal re-computation can effectively capture the dynamics of a closed-loop system. Furthermore, the results demonstrate that ANTIFRAGILE outperforms both ROBUST and OPTIMAL concerning transient parameters and steady-state error.

In the course of our experiments, we observed that the OPTIMAL control system attempted to optimize the performance metrics over the specified period, but was highly sensitive to the parameters of the oscillator-based network model. In contrast, the ROBUST control system aimed to optimize the stability and quality of the response in the presence of mild levels of uncertainty (i.e., disruptions) in the oscillator-based model. The ANTIFRAGILE employs variable structure control to address both structured and unstructured uncertainty in the model by synthesizing a second-order effects-aware control that propels the system towards the desired antifragile region of the state space while ensuring stability.

4.1.6 Conclusion

The optimization of traffic control is a complex process, given the inherent uncertainty involved. Modelling traffic dynamics is a fundamental aspect of traffic control.

To capture the periodic nature of traffic, we propose an instantiation of an antifragile control for modelling urban road traffic. The proposed control employs a network of oscillators to capture the spatial and temporal interactions among different intersections in a traffic network. To adaptively cope with unexpected traffic flow disruptions, the antifragile control employs a second-order attractor dynamics controller that enhances its capacity for adaptation towards global consensus in the event of significant disruptions and uncertainty. The control-theoretic design and analysis, together with the experiments on real-world data, provide quantitative evidence of the advantages of antifragile control over static, robust, and optimal signal control in terms of traffic quality metrics.

4.2 Antifragile Robotic Systems

This section discusses the design of Antifragile controllers for robotic systems, with a special focus on mobile robots. We first present the motivation for applying antifragility principles in robot control, and then present a practical implementation for a wheeled robot.

4.2.1 Trajectory Tracking for Wheeled Mobile Robots

Mobile robotics has sparked the control community's interest in the context of human-assistive applications. Such wheeled mobile robots are often characterized as nonholonomic mechanical systems. For many years, nonholonomic vehicle control has been a hotly debated research topic. This is due to at least two factors. On the one hand, nonholonomic wheeled robotic vehicles are an important and increasingly common mode of mobility. Previously only seen in research labs and factories, autonomous robotic vehicles are increasingly being adopted in everyday life (e.g. through car-platooning applications, as shown in autonomous mobility by [32], geriatronics applications of [33], or urban transportation services described by [34]).

Trajectory tracking control of nonholonomic mobile robots seeks to control a robot's motion to follow a specific time-varying trajectory. It is a basic motion control challenge that the robotics community has studied in great detail, proof of the pioneering works of [35–38], and, of course, our previous work in [39]. The tracking control problem is categorized as either kinematic or dynamic depending on whether the system is described by a kinematic or dynamic model.

Several researchers have investigated the kinematic tracking problem and developed various types of controllers. The seminal work of [40] addressed the trajectory-tracking problem by using the kinematic model of a wheeled mobile robot. In a more theoretical work [41] addressed both local and global tracking problems with exponential convergence utilizing time variable state feedback based on the backstepping approach. As we see, the kinematic tracking control problem for mobile robots has received much research, but the dynamic tracking control problem has only recently attracted attention.

The majority of the results on dynamic model-based tracking problems of nonholonomic systems are presented on the assumption that the system's kinematics are precisely understood and that uncertainties exist only in the dynamics. In practice, however, errors exist in both kinematics and dynamics. Typically, the reference trajectory is derived by employing a reference (virtual) robot; hence, the reference trajectory takes into account all kinematic restrictions implicitly. The majority of the control inputs are produced using a mix of feed-forward inputs estimated from the reference trajectory and feedback control rule, as shown in the work of [35, 36, 42]. In the same context, the work of [40, 43], and, of course, [43] pioneered Lyapunov stable time-varying state-tracking control rules, in which the system equations are linearized about the reference. The controller parameters are calculated by defining the desired parameters of the characteristic polynomial. A nonzero motion condition is required for stability to the reference trajectory. Along the same lines, the work of [44] introduced a discontinuous stabilizing controller for mobile robots with nonholonomic restrictions, in which the robot's state asymptotically converges to the goal configuration with a smooth trajectory.

More in line with our uncertainty handling approach, the work of [45] developed a tracking problem for a mobile robot to follow a virtual target vehicle that moves precisely along a path with a given velocity. To minimize wheel slippage or mechanical damage during navigation, the driving velocity control rule was created based on bang-bang control while taking the acceleration boundaries of the driving wheels and the robot's dynamic restrictions into account. Also relevant, the work of [37] designed a tracking controller for a differential drive mobile robot that is sensitive to wheel slip and external stresses using dynamic modelling.

As we see, many researchers have employed various nonlinear control strategies when dealing with system disturbances, operating uncertainty, and unknown dynamic characteristics. Similar in nature to one component of our approach, and used to tackle the tracking control problem for mobile robots, the pioneering work of [46], the work of [47–50] employed sliding mode motion control techniques, robust adaptive control techniques, and higher order sliding mode techniques, respectively. More precisely, they proposed variable structure control approaches, using sliding mode control for the trajectory tracking issue for mobile robots in the presence of disturbances that violate the nonholonomic constraints. Finally, [51] and the excellent work of [52] established a model-based control design technique for the kinematic model with a nonholonomic mobile robot in the presence of input saturation that deals with global stabilization and global tracking control which yielded comparable results to the non-parametric adaptive control approach of [53] and the neural network robust control approach of [54].

Although well rooted in the robotic control field, among the previously presented works, the current study covers a unique control approach for nonholonomic vehicles, namely antifragile control, and more precisely trajectory tracking in the face of uncertainty, volatility, and unpredictability. This approach goes beyond our initial explorations in [55] and tries to propose to the community a novel perspective on robot motion control, namely antifragile control.

4.2.1.1 Fragility-Robustness-Antifragility Spectrum in Robot Control

Trajectory tracking requires a task planning step. At the planning level, autonomous robot vehicles produce their judgments that determine how to operate the vehicle actuators and cause the vehicle to move, as shown in the work of [56]. The challenge with motion planning and control is that the motion constraints of any actuators involved or the vehicle platform itself must be considered, as formally described in [57]. This is especially relevant for wheeled mobile robots, which are constrained by nonholonomic constraints. This means that a vehicle travelling on a surface may have three degrees of freedom: two degrees of translation and one degree of rotation. As a result, the equations of motion that describe vehicle dynamics are non–integrable, making the problem significantly more complex to solve. This also implies that wheeled mobile robots are under-actuated. In other terms, the system's number of control inputs is smaller than the number of degrees of freedom in its configuration space. Additionally, the uncertainty related to the travelling surface, sensors, and actuator faults is an additional dimension to consider in control design.

The main goal of this study is to introduce the application of antifragile control to mobile robot trajectory tracking control under uncertainty, volatility, and variability of the operating environment and the robot's sensors and actuators. According to [99], antifragility is a feature of a system that allows it to benefit from uncertainty, unpredictability, and volatility, in contrast to fragility. The reaction of an antifragile system to external perturbations is beyond robust and resilient such that mild stresses can increase the system's future response by adding a significant anticipation component. In this work, we propose an alternative control mechanism, based on the antifragile control framework introduced by [15], further refined and extended in [58] and built on top of the principles in the seminal work of [14].

To instantiate the antifragile control framework for robot trajectory tracking control, we need to define the Fragility-robustness-antifragility spectrum. To guide the reader with intuition on the benefits of antifragile control, we consider a simple depiction of how various types of controllers would perform in the presence of gradually increasing uncertainty (e.g. wheel slippage, actuator fault, or sensor fault). We consider a hypothetical effect only for graphical purposes.

The purpose of Fig. 4.10 is to delineate, in a graphical and easy-to-grasp manner, the main concept of the proposed approach. Achieving an antifragile closed-loop control performance, that not only compensates for unexpected, increasingly strong disturbances but also gains from subsequent exposures, is the core motivation of our work. The actual implementation details follow in the next sections along with more intuitive aspects that strengthen this hypothetical depiction of the robot's response.

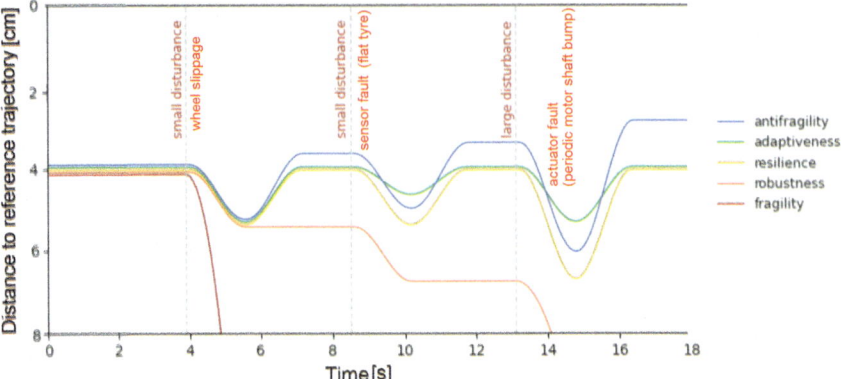

Fig. 4.10 Fragility-robustness-antifragility spectrum in robot trajectory tracking control. Uncertainty in a robot's motion can emerge from environmental parameters (e.g., wheel slippage), sensor faults (e.g., perceiving a continuous wheel radius decrease during operation akin to a flat tire), or actuator faults (e.g., a periodic eccentric mechanical motion of the DC motor shaft akin to a wheel bump). The possible closed-loop system responses are reflected in the actual displacement from the reference trajectory to track. We can see that with the increase in amplitude and timing (i.e., uncertainty and volatility) of the disturbance the system can compensate up to a point but, in contrast to antifragile control, cannot gain from the adverse events. It is important to note the reaction time and the amplitude of the response with respect to the occurrence and strength of the adverse event

4.2.2 Wheeled Robot Modelling and Control

In this section, we introduce the models and tools we employed in our study. In particular, we derive the mathematical model of a wheeled mobile robot and present basic techniques to design a trajectory-tracking controller.

4.2.2.1 Modelling a Nonholonomic Mobile Robot

This subsection provides an overview of the modeling of nonholonomic mobile robots for trajectory tracking. We remind that nonholonomic constraints make the motion perpendicular to the wheels impossible, requiring nontrivial control mechanisms in trajectory tracking.

We consider a differential drive wheelchair as depicted in Fig. 4.11. The notations for the reference systems and the kinematic quantities follow the standard conventions. More in detail, all kinematic quantities are defined in the local coordinate (reference) system XP_0Y, whereas the control and measurements will be mapped to the world reference system xOy. P_0 is the origin of the local coordinate system fixed at the middle point between the right and left driving wheels. The distance from P_0 to the centre of mass P_c is d. Each driving wheel has a radius r and the distance between wheels is $2b$. The heading angle of the robot is ϕ. Additionally, for

Fig. 4.11 Wheelchair-type mobile robot with differential drive considered in this chapter. Adapted with permission from [59]

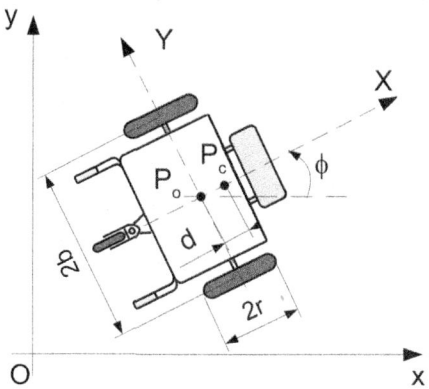

our robot, we assume that the velocity of P_0 must be in the direction of the axis of symmetry and the wheels must not skid (i.e., motion constraints).

A mobile robot system with an n-dimensional configuration space, generalized variables $(q_1, q_2, ..., q_n)$, and constraints may be expressed, following the work of [60], as following:

$$M(q)\ddot{q} + V_m(q,\dot{q})\dot{q} + F(\dot{q}) + G(q) + \tau_d = B(q)\tau - A^\top(q)\lambda, \quad (4.32)$$

where $M(q) \in R^{n \times n}$ is a symmetric positive definite inertia matrix of the robot, $V_m(q,\dot{q}) \in R^{n \times n}$ is the centripetal and Coriolis matrix, $F(\dot{q}) \in R^{n \times 1}$ describes the surface friction of the robot, $G(q) \in R^{n \times 1}$ is the gravity vector, τ_d describes the overall bounded unknown disturbances including unstructured unmodelled dynamics, $B(q) \in R^{n \times r}$ is the input gain matrix, $\tau \in R^{n \times 1}$ is the input vector of the robot, $A(q) \in R^{m \times n}$ is the constraints matrix, and $\lambda \in R^{m \times 1}$ is the vector of constraint forces acting upon the robot. The nonholonomic character of the mobile robot is associated with the notion that the robot's wheels roll without sliding. They are constrained by nonholonomic non–integrable equality requirements concerning velocity. In other words, the permissible velocity space has a lower dimension than the configuration space. This limitation can be expressed as $A(q)\dot{q} = 0$, where

$$A(q) = \begin{bmatrix} sin(\phi) & -cos(\phi) & d & 0 & 0 \\ cos(\phi) & sin(\phi) & b & -r & 0 \\ cos(\phi) & sin(\phi) & -b & 0 & -r \end{bmatrix}. \quad (4.33)$$

But, for control, the configuration of the mobile robot may be described using five generalized coordinates, $q = [x, y, \phi, \theta_r, \theta_l]^\top$, where (x, y) are the coordinates of the point P_0 (see Fig. 4.11), ϕ is the heading angle of the robot, and θ_r, θ_l are the angles of the right and the left driving wheels, respectively. If we let $S(q)$ be a full rank matrix formed by a set of smooth and linearly independent vectors such that $S^\top(q)A^\top(q) = 0$ then it is easy to verify that $S(q)$ is given by

$$S(q) = \begin{bmatrix} \frac{r}{2b}(bcos(\phi) - dsin(\phi)) & \frac{r}{2b}(bcos(\phi) + dsin(\phi)) \\ \frac{r}{2b}(bsin(\phi) + dcos(\phi)) & \frac{r}{2b}(bsin(\phi) - dcos(\phi)) \\ \frac{r}{2b} & -\frac{r}{2b} \\ 1 & 0 \\ 0 & 1 \end{bmatrix}. \tag{4.34}$$

Then according to Eq. (4.32) and the fact that $S^\top(q)A^\top(q) = 0$, it is straightforward to find that

$$\dot{q} = S(q)\omega, \tag{4.35}$$

where $\omega = [\omega_r, \omega_l]^\top$ is the vector of angular velocities of the right and left wheel, respectively. Equation (4.35) is the kinematic model of the robot. For the interested reader, differentiating Eq. (4.35) and substituting the result in Eq. (4.32), and then multiplying by $S^\top$ we can, of course, eliminate the constraint matrix $A^\top(q)\lambda$ and obtain the dynamic model of the robot in the form

$$\bar{M}(q)\dot{\omega} + \bar{V}_m(q, \dot{q})\omega = \bar{B}(q)\tau, \tag{4.36}$$

where $\bar{M} = S^\top MS$, $\bar{V}_m = S^\top(M\dot{S} + V_m S)$ and

$$\bar{M}(q) = \begin{bmatrix} \frac{r^2}{4b^2}(mb^2 + I) + I_w & \frac{r^2}{4b^2}(mb^2 - I) \\ \frac{r^2}{4b^2}(mb^2 - I) & \frac{r^2}{4b^2}(mb^2 + I) + I_w \end{bmatrix}, \tag{4.37}$$

$$\bar{V}_m(q) = \begin{bmatrix} 0 & \frac{r^2}{2b}m_c d\dot{\phi} \\ -\frac{r^2}{2b}m_c d\dot{\phi} & 0 \end{bmatrix}, \bar{B} = \begin{bmatrix} 1 & 0 \\ 0 & 1 \end{bmatrix}, \text{ and } \tau = \begin{bmatrix} \tau_r \\ \tau_l \end{bmatrix}, \tag{4.38}$$

where m_c is the mass of the robot's body and m_w is the mass of a driving wheel plus its associated motor, I, I_w are the moments of inertia of the body around the vertical axis through P_c and the driving wheel (with a motor) about the wheel axis, respectively.

When considering the dynamic model in Eq. (4.36), accurate knowledge about the parameter values of the mobile robot dynamics is nearly impossible to obtain in practice. If we consider that these parameters are also time-varying, the problem becomes even more complicated. It was originally proven in the work of [61] that a continuous (smooth) time-invariant pure state feedback rule, resulting from a violation of Brocketts' necessary condition for stability, cannot stabilize a nonholonomic system to a single equilibrium point. Furthermore, a wheeled mobile robot is only locally controllable over short time intervals, according to [61], and it is a controllable system independent of the nature of the nonholonomic constraints $A^\top(q)\lambda$ as shown by [62]. As a result, the control options are either (a) discontinuous time-invariant feedback laws or (b) continuous but time-variable non-linear feedback control laws applied to the model in Eq. (4.35). More precisely, for the controller design, we will use the explicit form of Eq. (4.35).

4.2 Antifragile Robotic Systems

$$\frac{d}{dt}\begin{bmatrix} x \\ y \\ \phi \\ \theta_r \\ \theta_l \end{bmatrix} = \begin{bmatrix} \frac{r}{2}\cos(\phi) & \frac{r}{2}\cos(\phi) \\ \frac{r}{2}\sin(\phi) & \frac{r}{2}\sin(\phi) \\ \frac{r}{2b} & -\frac{r}{2b} \\ 1 & 0 \\ 0 & 1 \end{bmatrix} \begin{bmatrix} \omega_r \\ \omega_l \end{bmatrix} \quad (4.39)$$

and given the known relation between the linear v and angular ω velocities of the robot and the individual wheel angular velocities ω_r, ω_l (i.e., knowing the wheel radius and distance between wheels), we can rewrite Eq. (4.39) as the ordinary form of a mobile robot with two actuated wheels in

$$\frac{d}{dt}\begin{bmatrix} x \\ y \\ \phi \end{bmatrix} = \begin{bmatrix} \cos(\phi) & 0 \\ \sin(\phi) & 0 \\ 0 & 1 \end{bmatrix} \begin{bmatrix} v \\ \omega \end{bmatrix} \quad (4.40)$$

Now, with all the modelling in place, we reiterate the objective of trajectory tracking as a control synthesis problem to compute the velocity of the robot such that its pose $P_r = [x_r, y_r, \phi_r]^\top$ follows a reference trajectory of the virtual robot $P_d = [x_d, y_d, \phi_d]^\top$.

4.2.2.2 Controlling a Nonholonomic Mobile Robot

The problem of trajectory tracking for fully actuated systems is now well known, and adequate solutions may be found in advanced nonlinear control textbooks. However, in the case of under-actuated vehicles, that is, vehicles with fewer actuators than state variables to be tracked, the problem is still a hotly debated research topic. Linearization and feedback linearization algorithms from [63, 64] have been developed, as have LyapunoV-based control laws, with representative designs in the work of [60, 65]. Independent of the synthesized control law, the trajectory tracking problem can be graphically formulated as shown in Fig. 4.12. The control algorithm needs to compensate for the heading Φ_e, lateral y_e, and longitudinal errors x_e and come closer to the virtual robot. The goal is to make the robot pose $P_r = [x_r, y_r, \phi_r]^\top$ follow a reference trajectory of the virtual robot $P_d = [x_d, y_d, \phi_d]^\top$.

Now, putting all elements together, we assume that a feasible desired trajectory for the mobile robot is pre-specified by a velocity planner from [38] and fed to a closed-loop control system that will ensure that the robot will correctly track the desired trajectory under a large class of disturbances. The motion of the robot, following the models above and the conventions in Fig. 4.12 is given by Eqs. (4.39) and (4.40).

$$\begin{cases} \dot{x}_r(t) = v_r(t)\cos(\phi_r(t)) \\ \dot{y}_r(t) = v_r(t)\sin(\phi_r(t)) \\ \dot{\phi}_r(t) = \omega_r \end{cases} \quad (4.41)$$

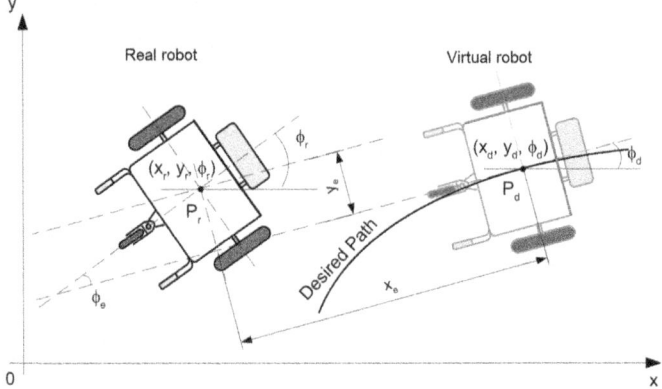

Fig. 4.12 Description of the mobile robot trajectory tracking problem. The real mobile robot tries to follow the desired path under time constraints. Adapted with permission from [59]

where x_r and y_r are the Cartesian coordinates of the geometric center of the mobile robot, v_r is the linear velocity of the robot, ϕ_r is the robot's heading angle, and ω_r is the angular velocity of the robot, respectively. The trajectory tracking errors can be described by the vector (x_e, y_e, ϕ_e) depicted in Fig. 4.12. The designed controller needs to generate a command vector (v_c, ω_c). Considering the ordinary form of the mobile robot in Eq. (4.41) the error vector, following the convention in Fig. 4.12, is given by Eq. (4.42).

$$\begin{bmatrix} x_e \\ y_e \\ \phi_e \end{bmatrix} = \begin{bmatrix} \cos(\phi_d) & \sin(\phi_d) & 0 \\ -\sin(\phi_d) & \cos(\phi_d) & 0 \\ 0 & 0 & 1 \end{bmatrix} \begin{bmatrix} x_r - x_d \\ y_r - y_d \\ \phi_r - \phi_d \end{bmatrix} \quad (4.42)$$

where the vector $[x_d, y_d, \phi_d]^\top$ is the virtual robot pose. The corresponding error derivatives are then given by Eq. (4.43).

$$\begin{cases} \dot{x}_e(t) = -v_d + v_r \cos(\phi_e) + y_e \omega_d \\ \dot{y}_e(t) = v_r \sin(\phi_e) - x_e \omega_d \\ \dot{\phi}_e(t) = \omega_r - \omega_d \end{cases} \quad (4.43)$$

where v_d and ω_d are the desired robot linear and angular velocities, respectively.

A final important component in the robot trajectory tracking control loop is the trajectory planner that generates the desired trajectory to track. Although mobile robots' motion planning has been extensively investigated in recent decades, the need to develop trajectories with minimal accelerations and jerks is not traceable in the technical literature. In our experiments, we used the excellent work of [66] to tackle velocity planning and provide suitable time sequences for use in interpolating curve

4.2 Antifragile Robotic Systems 77

planners. Using this approach allowed us to develop speed profiles (i.e., both linear and angular) that lead to trajectories that are comfortable for humans, as validated in the study of [38].

4.2.3 Antifragile Control of Nonholonomic Robots

This section is dedicated to introducing the mathematical apparatus of antifragile control, going from its theory and principles to the control synthesis for robot trajectory tracking under uncertainty. Figure 4.13 shows the synthesis of the antifragile controller, based on Eqs. (4.36), (4.39), and (4.34), respectively.

In the context of control systems, the generation of such behavior (i.e., induced antifragility) within a feedback control loop offers a distinctive design and synthesis methodology. (1) Redundant overcompensation may result in the system overshooting, thereby accumulating additional capacity and capability in anticipation; (2) Structure-variability can elicit stressors that carry intrinsic information, which emerges only in the context of system dynamics, the application of a high-frequency component introduces volatility and unpredictability. Furthermore, the interacting

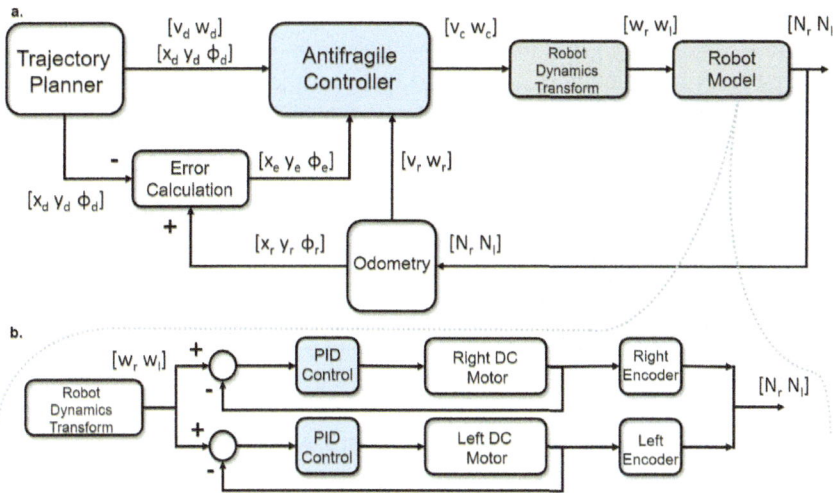

Fig. 4.13 Global control structure for antifragile control of mobile robot trajectory tracking. **a** The outer loop contains slower dynamics of pose correction based on the antifragile controller that uses the measured and reference pose vector to compute new linear and angular velocity values. **b** The inner loop contains faster dynamics of the two actuators (i.e., DC motors), which control the individual angular velocities of the wheels based on the motion of the motors measured through position encoders. Notations and conventions are consistent with Fig. 4.12 and the equations in the section. The Robot Dynamics Transform is the inverse kinematics transformation from $[v_c, w_c]$ to $[w_r, w_l]$

system's dynamics undergo a reduction in order when driven towards the desired antifragile operation region. This phenomenon can be explained by the preceding notions of redundancy, variability, and time-scale separation. We will now elucidate how these concepts relate to the robot trajectory control problem and discuss their potential applications in practice.

4.2.3.1 Control Synthesis

Previously, in [15, 58], we cast the control design in geometric control and Riemannian geometry objects, as formally explained in [16]. This approach enabled us to work in a coordinate-free environment, relying on the embedding of a manifold into a wider dynamical space. This allowed for simpler control law definitions that were adequate for manifolds with curvature. In this study, we consolidate this approach and reduce several previous assumptions while providing concrete control synthesis steps. Readers interested in further theoretical insights on casting the antifragile control theory in the Riemannian geometry framework may wish to refer to [15].

The objective is to design a controller that forces the robot to track a prescribed trajectory (i.e., a velocity-parametrized reference temporal evolution) with specific geometrical properties. Alternatively, the problem can be formulated to compute a control signal (i.e., reference linear and angular velocities) such that the robot dynamics state trajectory is confined to a desired dynamics, where the error vector is minimized (x_e, y_e, ϕ_e) is minimized. In other words, we want to drive the closed-loop system state evolution to a manifold such that the longitudinal x_e, the lateral error y_e, and the angular error ϕ_e are internally mutually coupled on the considered manifold leading to convergence of all three variables.

In our control synthesis, we decouple the two internal control loops in Fig. 4.13 (see the darkly shaded boxes termed Antifragile Control and PID control) to describe the specific design steps of (1) redundant overcompensation; (2) structure-variability; and (3) time scales separation for uncertainty isolation.

Time scale separation

Given the interactions between the two nested control loops (see Fig. 4.13 internal DC motor control loop and the outer position control loop), to handle uncertainty and high-frequency phenomena, we need to enable the closed-loop system to separate the time scales of the loops. A very useful tool for such interventions in closed-loop control is (singular) perturbation theory, initially proposed by [67] and further extended in [68]. In this context, the high-frequency dynamics are considered on a distinct timescale. This transformation is achieved by a dynamic alteration in the order of the controlled system as a parameter perturbation, which may be conceptualized as a parallel transport map on the Riemannian manifold of the system state trajectory.

The alteration in the controlled system dynamics is more pronounced than a typical perturbation to which the system is subjected, thereby characterizing it as a singular perturbation. The principal rationale for employing this methodology in the context of our antifragile control design is the observation that these "parasitic",

4.2 Antifragile Robotic Systems

high-frequency phenomena possess the capacity to develop resilience in response to significant fluctuations in the robot's operational parameters, including instances of wheel slippage, tire punctures, and shaft bending. The objective of this section is to familiarize the reader with the concept of time scale separation through singular perturbation theory as a component of antifragile control and to illustrate its practical application in controller synthesis.

The fundamental concept of time scale separation is to identify and quantify the dominant phenomena dynamics, as well as to isolate and analize the stressors. This is typically achieved through the utilization of either "outer" series expansions or "inner" boundary layer expansions, as proposed in [69] and graphically depicted in Fig. 4.14.

In the context of singularly perturbed dynamical systems, the concept of fast variation zones offers a valuable avenue for insight. These zones, which may be identified

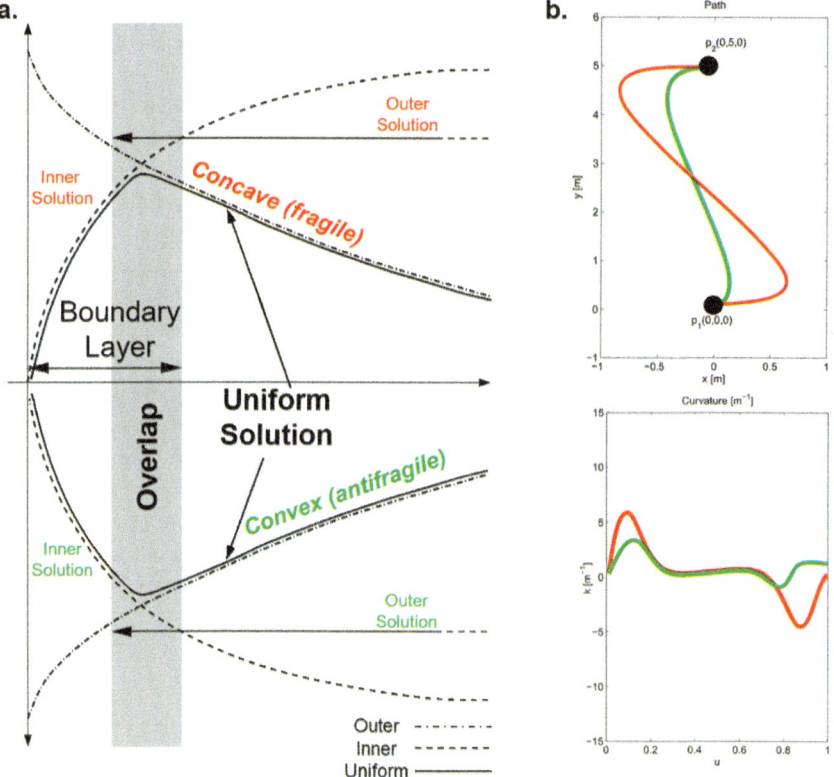

Fig. 4.14 Singular perturbation for time scale separation in antifragile controller synthesis. Using boundary layers and matched asymptotic expansions to probe antifragile behavior. **a** Generic depiction of a boundary layer and the types of solutions in singularly perturbed dynamic systems. **b** Mapping the boundary layers and shape (convexity/concavity) of the solution to robot velocity planning akin to the desired dynamics to track in the presence of uncertainty

in the solution or its derivatives, are commonly referred to as "layers" and are often situated in proximity to the domain border, as shown in Fig. 4.14. The construction of a solution to a differential equation of a dynamical system is a multi-stage process. Initially, the locations of layers must be determined, distinguishing between border and internal layers. Subsequently, asymptotic approximations to the solution must be obtained in different regions, corresponding to distinct differentiated limits in the equations. Finally, a uniformly valid solution must be produced throughout the entire domain, as described in [70] and depicted in Fig. 4.14a. Inner solutions are identified for the layers, whereas outer solutions are obtained for the regular distinguished limits. The uniform solution is characterized by the curvature (i.e., 2nd order effect) of the overlapping region between the inner and the outer layer. Notably, this can be utilized in our design to define fragile and antifragile control regimes contingent on the curvature in the overlap region of the solutions (i.e., attractors/solutions in the system's state space). As depicted in Fig. 4.14a, we define the antifragile region as the convex region of the solution curve. Hence, the closed-loop system response is antifragile if the curvature is negative; otherwise, it is fragile (see Fig. 4.14a).

To elaborate further, in the context of robot trajectory tracking control, the reference trajectory can be defined as a path, which is an explicit function of time (see Fig. 4.14b). To achieve a smooth robot movement, the trajectory must be twice differentiable, thus ensuring a continuous velocity and acceleration. Consequently, curve fitting constitutes an integral part of trajectory planning. The most effective method, as demonstrated in [59], is the use of piece-wise quintic polynomials, also known as quintic splines. The quintic splines are optimal for use in this context, as they facilitate continuity in position, heading, curvature, velocity, and acceleration. In our experiments, we used the method of [59] to obtain longitudinal and angular velocity profiles ($v(t)$ and $\omega(t)$). The profile must be compatible with the characteristics of the robot actuators, specifically DC motor regimes, and the assigned total path length. Furthermore, it must comply with the requisite human comfort travel parameters.

Let's now intuitively explore the mapping between the shape of the solution of the dynamics of the robot, depicted in Fig. 4.14a, to the actual control signals $v(t)$ and $\omega(t)$ needed to track the prescribed trajectory, depicted in Fig. 4.14b upper panel. This will support the need for such a design component in inducing an antifragile behavior. Figure 4.14a depicts the solution space of the planned trajectory of the robot given the explicit expression of curvature $k_{i,i+1}$ between consecutive points i and $i+1$ as

$$k_{i,i+1}(t) = \frac{\dot{x}_{i,i+1}\ddot{y}_{i,i+1} - \ddot{x}_{i,i+1}\dot{y}_{i,i+1}}{\dot{x}_{i,i+1}^2 + \dot{y}_{i,i+1}^2}, \qquad (4.44)$$

whose solution must be compatible, as mentioned above, with the DC motor regimes, and assigned total path length that must comply with human comfort travel. Based on the solutions of Eq. (4.44) and their sign (i.e., antifragile control signals for tracking if the curvature is negative, otherwise fragile), the computed signals $v(t)$ and $\omega(t)$ to move the robot from point p_1 to point p_2 will generate two possible paths, *fragile* (red) and *antifragile* green (Fig. 4.14). Both fragile (red) and antifragile (green) are feasible solutions. The antifragile solution will limit the curvature and, implicitly, the magnitude of the control input $v(t)$ and $\omega(t)$. This reduces the stress on the robot's actuators and ensures greater resilience in the event of uncertainty while maintaining comfort. The delicate path and curvature will exhibit a more pronounced curvature variation at the outset and conclusion of the spline, respectively, which may potentially compromise resilience in the event of, for instance, wheel slippage or mechanical damage during navigation.

As demonstrated in the preceding antifragile control instantiation of Axenie (2022), oscillators are periodic dynamical systems exhibiting "rapid" (inner layers) and "slow" (outer) dynamics (see Fig. 4.14). It was then postulated that a uniformly valid solution may be produced by the asymptotic matching of the inner and outer solutions, which is based on the fundamental premise that the various solution forms overlap at some recognizable location, typically provided by the curvature.

Redundant overcompensation

In the following, we revisit the core idea of time scale separation, graphically depicted in Fig. 4.14. Let us consider a more general form of the robot model as

$$\begin{cases} \dot{x} = f(x, z, \varepsilon, t), \ x(t_0) = x^0, \ x \in \mathbb{R}^n \\ \varepsilon \dot{z} = g(x, z, \varepsilon, t), \ z(t_0) = z^0, \ z \in \mathbb{R}^m \end{cases}, \quad (4.45)$$

where f, g are continuous differentiable functions of x, z, ε, t, basically accounting for the robot model in Eq. (4.39). The scalar ε quantifies all small parameters of the system (i.e., I_m, I_w, etc.), which in the antifragile control framework are termed as stressors for capacity build-up. Furthermore, if we consider T_1 and T_2 two small time constants of the same order of magnitude, we can assume that they can be taken as ε and have, let's say $T_1 = \varepsilon$ and $T_2 = \alpha\varepsilon$, where $\alpha = T_2/T_1$ is a known constant. Now, if we set $\varepsilon = 0$ in Eq. (4.45), the dimension of the state space of the system reduces from $n + m$ to n because the second equation degenerates into the transcendental equation $0 = g(\hat{x}, \hat{z}, 0, t)$ where z can rapidly converge to a root of the transcendental equation due to its velocity $\dot{z} = g/\varepsilon$, which can be high if ε is small.

From a more intuitive perspective, the model in Eq. (4.45) is a reduced-order modelling technique, which allows us to convert the robot's dynamics simplification (reduction) into a parameter perturbation, called "singular". The solutions of the "slow" dynamics $x(t, \varepsilon)$ and the "fast" dynamics $z(t, \varepsilon)$ of the singularly perturbed system in Eq. (4.45) consist of a fast boundary layer and a slow quasi-steady-state, as shown in Fig. 4.14. From the Riemannian geometry perspective of antifragile control, there exists a manifold M_ε depending on ε that can be defined in the space $n + m$ of x and z such that $M_\varepsilon : z = \phi(x, \varepsilon)$. This reduces the dimension of the state space by restricting it to remain on M_ε manifold. This integrates nicely with the variable structure systems formalism, which is also a fundamental design component in induced antifragile control.

In light of the formalism introduced in the preceding two sections, we now present the explicit design and implementation of the time scale separation and redundant overcompensation of the mobile robot antifragile controller. We will consider the inner control loop of the robot actuators in Fig. 4.13b. In this section, we examine the

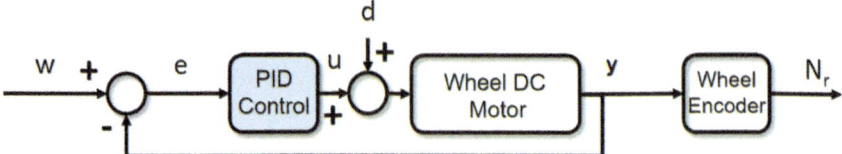

Fig. 4.15 Singular perturbation approach for the synthesis of an antifragile low-level inner loop controller based on the proportional-integral-derivative (PID) formulation. The proposed methodology employs time scale separation to facilitate redundant overcompensation of the low-level actuator control loop

synthesis of the two Proportional Integral Derivative (PID) controllers for closed-loop DC motor control, as illustrated in Fig. 4.15.

We begin with the standard formulation of the PID controllers and a simplified model of the DC motor, as typically found on robotic wheelchairs. To focus on the singular perturbation design, we will formulate the problem in the Laplace domain of complex frequency s. For the sake of simplicity, we will only work with algebraic forms of the control law. We will revisit the time domain to describe the actual time scale separation later on.

Referring to Fig. 4.15 and using the typical control theory conventions, the transfer function of the PID controller $u(s)$ is $u(s)/e(s) = K_P(1 + K_D s + K_I \frac{1}{s})$ where the error $e(s) = w(s) - y(s)$. Furthermore, we consider the DC motor model

$$\begin{cases} J\dot{\omega} = ki \\ L\dot{i} = -k\omega - Ri + u \end{cases} \quad (4.46)$$

where i, u, R and L are the DC motor's armature current, voltage, resistance, and inductance respectively, J is the moment of inertia, ω is the DC motor shaft angular speed, and ki and $k\omega$ are the torque and the back e.m.f. developed with constant excitation flux. In almost all well-designed motors, L is small and may serve as our parameter ε. This maps to our generic formulation in Eq. (4.45), where $\omega = x$, $i = z$ and the model in Eq. (4.46) has the conventional form in Eq. (4.45) when $R \neq 0$. We address the model reduction problem by ignoring the inductance L and solving $-k\omega - Ri + u = 0$ to obtain the value of the current $i = \frac{u - k\omega}{R}$ which we then substitute in Eq. (4.46) to obtain the first-order model of the DC motor in the form $J\dot{\omega} = -\frac{k^2}{R}\omega + \frac{k}{R}u$. This accounts for finding the manifold M_ε and restricting the DC motor dynamics to remain on it.

It is important to highlight in our design that the overall effect of the PID control is significantly slower than that of the proportional and derivative components. The singular perturbation theory supports us in the task of rewriting the control law u given the fact that K_P and K_D offer speed and stability of the system, respectively, whereas K_I reduces the error e to zero. Assuming that K_I is the same order of magnitude with ε, that is, $K_I = \varepsilon \hat{K}_I$, and changing notation as to $k_1 = K_P$, $k_2 = K_P K_D$, and $k_3 = K_P \hat{K}_I$, the control law is $u(s) = (k_1 + k_2 s + \varepsilon k_3 \frac{1}{s})e(s)$.

4.2 Antifragile Robotic Systems

The state variables in the "fast" timescale $\tau = \frac{t}{\varepsilon}$ are $e_1 = e$, $e_2 = \frac{de}{d\tau}$, $e_3 = \varepsilon \int_0^\tau e \, d\sigma$. Therefore, the state representation of the low-level actuator control loop is described by

$$\begin{cases} \frac{de_1}{d\tau} = e_2 \\ \frac{de_2}{d\tau} = -k_1 e_1 - k_2 e_2 - k_3 e_3 - d \\ \frac{de_3}{d\tau} = \varepsilon e_1 \end{cases} \quad (4.47)$$

If we rewrite the system in the "slow" time variable $t = \varepsilon \tau$ and we identify, following the generic formulation in Eq. (4.45), selecting $e_1 = z_1, e_2 = z_2$ as the fast variables, and $e_3 = x$, as the slow variable respectively, then we have the following formulation of the closed-loop system

$$\begin{cases} \dot{x} = z_1 \\ \varepsilon \dot{z}_1 = z_2 \\ \varepsilon \dot{z}_2 = -k_1 e_1 - k_2 e_2 - k_3 e_3 - d \end{cases} \quad (4.48)$$

The fast variables will enable the robot to respond effectively to uncertain events, such as wheel slippage, flat tires and DC motor actuator shaft bending. To achieve this, we must select appropriate values for k_1 and k_2. This will ensure that the system matrix of Eq. 4.48 is Hurwitzian in the PD part of the motor controller, that is

$$\text{Re}\left\{\lambda\left(\begin{bmatrix} 0 & 1 \\ -k_1 & -k_2 \end{bmatrix}\right)\right\} < 0. \quad (4.49)$$

Finally, the integral (I) component of the reduced order (dominant) PID controller is obtained by setting the fast variables to $z_1 = -\frac{k_3 e_3 + d}{k_1}$ and $z_2 = 0$ (recall that $e_1 = z_1$ and $e_2 = z_2$) such that the state evolution is $\dot{x} = -\frac{k_3}{k_1} x - \frac{d}{k_1}$, given the previous notation $\omega = x, i = z$. Then, the boundary layer system (the overlap region in Fig. 4.14 where the uniform solution's convexity can be probed) is given by simplified dynamics as follows

$$\begin{cases} \frac{d\hat{z}_1}{d\tau} = \hat{z}_2 \\ \frac{d\hat{z}_2}{d\tau} = -k_1(\hat{z}_1 - \dot{x}) - k_2 \hat{z}_2 - k_3 x - d = -k_1 \hat{z}_1 - k_2 \hat{z}_2 \end{cases} \quad (4.50)$$

Structure variability

As discussed in the preceding two subsections, the effects of time scale separation and low-level redundant overcompensation using singular perturbation theory can be combined with another component of the antifragile control synthesis, namely structure variability. This concept is already recognized within the community, but the challenge in control systems design lies in operating under conditions of high uncertainty. Variable structure control (VSC) systems offer a highly effective tool for handling uncertainty in closed-loop systems, as demonstrated in the seminal work of [71].

It is typically thought that the best way to withstand uncertainty is through "brute force". However, it is also known that any strictly enforced equality removes one "uncertainty dimension". Therefore, there is always a cost to be paid for precisely attaining the control goal, as formally described in [18]. VSC provides an appropriate tool for controller design, offering a robust response to minor deviations from a specified constraint. In practice, VSC is typically implemented through sliding mode control, which was introduced by [72] and further extended in [73].

Sliding mode controllers guarantee that the maximum deviation from a constraint is directly proportional to the time interval between the system's observations. The design follows a model reduction principle based on singular perturbation theory. Consequently, we have developed an antifragile control system with a unified framework that allows for redundant overcompensation, time-scale separation, and variable structure control.

The advantages of using VSC and sliding mode in our antifragile control design for robot trajectory tracking are listed below:

- The motion equation of the sliding mode (i.e., the prescribed dynamics), as set out in Ref. [18], can be designed to be linear and lower-order, despite the nonlinear effects of robot dynamics and uncertainties. (see Fig. 4.16b).
- The sliding manifold (i.e., both a place and the dynamics of the closed-loop robot control) is not dependent on the robot model. Rather, it is determined by parameters selected by the designer, as outlined in [71]. In our particular context, we can ensure that the robot follows the desired trajectory in the antifragile region of the prescribed trajectory planning solution, taking into account the constraints related to the actuators, the environment and the comfort of the operator (see Fig. 4.14).
- Once the sliding motion occurs (i.e., the system dynamics are on the manifold), the robot motion in trajectory tracking exhibits invariant properties that render the motion independent of certain system parameter variations, uncertainty, and disturbances, as detailed in [72]. Hence, the system performance can be completely determined by the dynamics of the sliding manifold, as depicted in Fig. 4.16.

To facilitate comprehension of the rationale behind the utilisation of VSC for antifragile control synthesis, we present a straightforward graphical illustration in Fig. 4.16. To address the robot trajectory tracking problem, it is necessary to implement an antifragile controller that can provide a suitable control signal. This signal should comprise a pair of longitudinal velocity, v, and angular velocity, ω, which will enable the path from the origin to the destination (see Fig. 4.16a) to be tracked following the specified time constraints. Furthermore, the controller must be capable of functioning effectively in the presence of uncertainty regarding the driving surface, potential actuator failures and increased comfort (i.e., minimal curvature). This is accomplished through the appropriate synthesis of the control law, which is constructed with a redundant over-compensation capacity to address uncertainty regarding the driving surface and potential actuator failures (see the initial conditions of the system dynamics converging from the green manifold to the induced antifragile dynamics illustrated in Fig. 4.16b). When the system is initiated from a fragile

4.2 Antifragile Robotic Systems

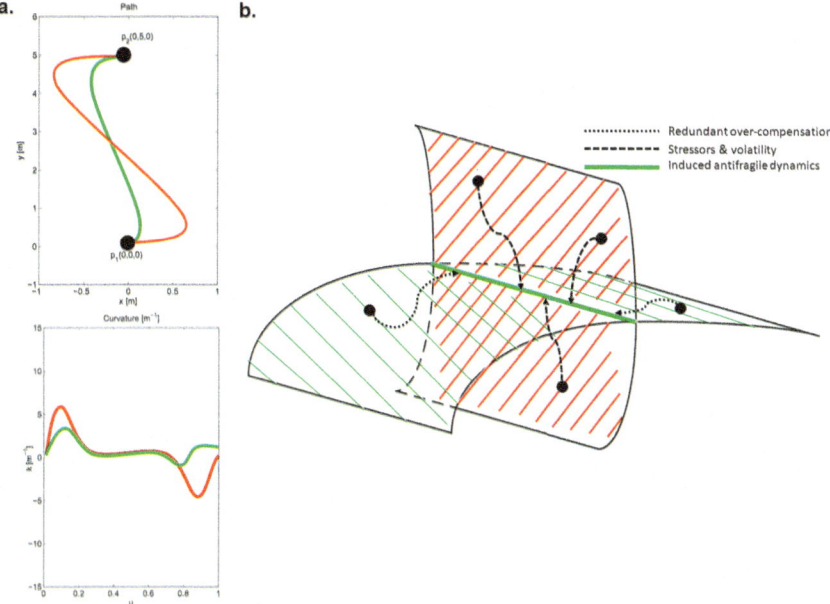

Fig. 4.16 The objective of this study is to investigate the potential of variable structure control through sliding modes in the context of antifragile control synthesis. To this end, we consider a robot trajectory tracking control problem. **a** In particular, we examine the ability to track a prescribed trajectory while accounting for time constraints, uncertainty, and comfort constraints (minimal curvature). The fragile dynamics (red) provide a feasible trajectory tracking solution with a control signal that reduces comfort. The antifragile control signal (green) reaches an increased level of comfort through a trajectory tracking performance that is both feasible and of higher quality, with a lower level of error. **b** Closed-loop system dynamics for fragile versus antifragile behaviors. The induced antifragile control manages to drive the system's dynamics towards the antifragile region (green thick intersecting line of the red and green manifolds), benefiting from redundant over-compensation, stressors, and volatility

region of the solution space (see red manifold in Fig. 4.16b), the controller responds to stressors and volatility (i.e., increased curvature of the trajectory) by utilizing the provided 'inertia' to converge to the induced antifragile dynamics.

To achieve the desired dynamics under the aforementioned antifragile control, we will now proceed to outline the essential design stages of a VSC, with a particular focus on the aspects pertinent to our problem.

Sliding manifold selection

In this control design step, the objective is to select a sliding manifold with a lower order than that of the system, ensuring that the system performance is achieved during the sliding motion. The choice of the sliding manifold is highly dependent on the problem at hand; therefore, we will motivate the selection for the trajectory tracking problem we consider. We begin with the canonical form of a sliding manifold, s, which depends on the system state dynamics, x. For a general dynamical system formulation, see Eqs. 4.45 and (4.51).

$$\dot{s} = \frac{\partial s}{\partial x}\dot{x} = \frac{\partial x}{\partial x}s(x) = \lambda_1 x_1 + \lambda_2 x_2 + \ldots + x_n = 0 \quad (4.51)$$

where the coefficients λ_i in $\dot{s}$ define the desired characteristics of the sliding mode, that is, the characteristics of the closed-loop system after the manifold reaching phase, as broadly described in [72]. Finding these parameters is typically formulated as an optimization problem and solved using linear programming techniques (e.g., Linear Quadratic (LQ) approach), as shown in [74]. Here, a criteria for a second order system $J = \int_{t_s}^{\infty}(x_1^\top Q_{11}x_2 + 2x_1^\top Q_{12}x_2 + x_2^\top Q_2 x_2)dt$ was minimized to get the optimal sliding manifold. Considering $Q_{12} = 0$ then the optimal control $x_2 = -Q_{22}^{-1}A_{12}^\top P x_1 = -kx_1$ where P is a p.d. matrix solution of the Ricatti equation $A_{11}^\top P + PA_{11} - PA_{12}Q_{22}^{-1}A_{12}^\top P = -Q_{11}$ where A is the input matrix of the system. The switching function is obtained by simply considering $s(x) = kx_1 + x_2 = [Q_{22}^{-1}A_{12}^\top P, I]x$.

In this instance, the objective is to select a sliding manifold that ensures mutual convergence of the longitudinal error (x_e), the lateral error (y_e), and the angular error (ϕ_e), which are internally coupled. This is achieved by taking into account the robot error in the outer loop (see Fig. 4.12)

$$\begin{bmatrix} x_e \\ y_e \\ \phi_e \end{bmatrix} = \begin{bmatrix} \cos(\phi_d) & \sin(\phi_d) & 0 \\ -\sin(\phi_d) & \cos(\phi_d) & 0 \\ 0 & 0 & 1 \end{bmatrix} \begin{bmatrix} x_r - x_d \\ y_r - y_d \\ \phi_r - \phi_d \end{bmatrix}, \quad (4.52)$$

where the vector $[x_d, y_d, \phi_d]^\top$ is the virtual robot pose. The corresponding error derivatives are then given by

$$\begin{cases} \dot{x}_e(t) = -v_d + v_r \cos(\phi_e) + y_e \omega_d \\ \dot{y}_e(t) = v_r \sin(\phi_e) - x_e \omega_d \\ \dot{\phi}_e(t) = \omega_r - \omega_d \end{cases}. \quad (4.53)$$

The sliding manifolds we chose for the robot trajectory tracking are

$$\begin{cases} s_1 = \dot{x}_e + \lambda_1 x_e \\ s_2 = \dot{y}_e + \lambda_2 y_e + \lambda_0 \mathrm{sgn}(y_e)\phi_e \end{cases} \quad (4.54)$$

with $\lambda_0, \lambda_1, \lambda_2 > 0$. Interestingly, if s_1 converges to 0, then x_e converges to 0. Additionally, if s_2 converges to 0, then at steady state $\dot{y}_e = -\lambda_2 y_e - \lambda_0 \mathrm{sgn}(y_e)\phi_e$. Here, for negative lateral error $y_e < 0$ then $\dot{y}_e > 0$ if and only if $\lambda_0 < \lambda_2 \frac{|y_e|}{|\phi_e|}$ and for a positive lateral error $y_e > 0$ then $\dot{y}_e < 0$ if and only if $\lambda_0 < \lambda_2 \frac{|y_e|}{|\phi_e|}$.

Control law design

In this phase, our objective is to design a switched feedback control law that meets the reaching condition (see Fig. 4.16b) and guides the system trajectory to

4.2 Antifragile Robotic Systems

the manifold within a defined time frame and subsequently maintains it there. In this study, we examine Gao's reaching law, introduced in [75] that employs the differential equation $\dot{s} = -Q\text{sgn}(s) - Ph(s)$, where $Q = \text{diag}[q_1, q_2, \ldots, q_n]$ with $q_i > 0, i = 1, \ldots, n$; $P = \text{diag}[p_1, p_2, \ldots, p_n]$, with $p_i > 0, i = 1, \ldots, n$; $\text{sgn}(s) = [\text{sgn}(s_1), \text{sgn}(s_2), \ldots, \text{sgn}(s_m)]^\top$; $h(s) = [h_1(s_1), h_2(s_2), \ldots, h_m(s_m)]^\top$; and $s_i h_i(s) > 0$ with $h_i(0) = 0$. The reaching time for x to move from an initial state to the switching manifold s_i is finite and given by

$$T_i = \frac{1}{p_i} \ln \frac{p_i |s_i| + q_i}{q_i} \tag{4.55}$$

Now, having the reaching law equation, we can determine the control law u that drives the robot on the prescribed trajectory for tracking. In our case, the control law is obtained by computing the time derivative (i.e., the velocity) of $s(x)$ along the reaching mode trajectory (see Fig. 4.16b) as $\dot{s} = \frac{\partial s}{\partial x}(A(x) + B(x)u) = -Q\text{sgn}(s) - Ph(s)$ where, in the generic form, A is the state transformation matrix and B is the control input gain matrix. We then have the control law given by $u = -(\frac{\partial s}{\partial x} A(x) + Q\text{sgn}(s) + Ph(s))(\frac{\partial s}{\partial x} B(x))^\top$. In this case, the resulting sliding mode is not preassigned but rather follows the natural state trajectory of a first-order switching scheme, as shown in [76]. Of course, the switching takes place depending on the location in the state space of the initial state, as shown in Fig. 4.16.

In our particular case, we choose the control law u as

$$\dot{s} = -Qs - P\text{sgn}(s) \tag{4.56}$$

with $P, Q > 0$. Opposite to the approach of [76], we use the proportional term $-Qs$ instead of the $\text{sgn}(s)$ to force the system's state to approach the switching manifold faster when $\dot{s}$ is large, while the discontinuous (magnitude) component is given by $h(s) = \text{sgn}(s)$ in the second term (i.e., the constant rate reaching). Now, given the ordinary form for control of the mobile robot

$$\frac{d}{dt}\begin{bmatrix} x \\ y \\ \phi \end{bmatrix} = \begin{bmatrix} \cos(\phi) & 0 \\ \sin(\phi) & 0 \\ 0 & 1 \end{bmatrix} \begin{bmatrix} v \\ \omega \end{bmatrix} \tag{4.57}$$

and the derivative of the manifold Eq. (4.54) as

$$\begin{cases} \dot{s}_1 = \ddot{x}_e + \lambda_1 \dot{x}_e \\ \dot{s}_2 = \ddot{y}_e + \lambda_2 \dot{y}_e + \lambda_0 \text{sgn}(y_e)\dot{\phi}_e \end{cases} \tag{4.58}$$

we perform simple mathematical manipulations to obtain the control law $u = [v_c, \omega_c]^\top$ where the linear acceleration is

$$\dot{v}_c = \frac{1}{\cos(\phi_e)}(-Q_1 s_1 - P_1 \text{sgn}(s_1) - \lambda_1 \dot{x}_e - \dot{\omega}_d y_e - \omega_d \dot{y}_e + v_r \dot{\phi}_e \sin(\phi_e) + \dot{v}_d), \tag{4.59}$$

and the angular velocity is

$$\omega_c = \frac{1}{v_e \cos(\phi_e) + \lambda_0 \mathrm{sgn}(y_e)} (-Q_2 s_2 - P_2 \mathrm{sgn}(s_2) - \lambda_2 \dot{y}_e - \dot{v}_r \sin(\phi_e) + \dot{\omega}_d x_e + \omega_d \dot{x}_e). \tag{4.60}$$

Note that the sign function $\mathrm{sgn}(\cdot)$ in the control signals can be replaced in the practical implementation by the saturation function $\mathrm{sat}(\cdot)$ with thresholds to reduce the chattering phenomenon. Now, let us define the Lyapunov function candidate

$$V = \frac{1}{2} s^\top s. \tag{4.61}$$

The time derivative $\dot{V}$ is given by

$$\dot{V} = s_1 \dot{s}_1 + s_2 \dot{s}_2 = s_1(-Q_1 s_1 - P_1 \mathrm{sgn}(s_1)) + s_2(-Q_2 s_2 - P_2 \mathrm{sgn}(s_2)) \tag{4.62}$$

or in a shorter form

$$\dot{V} = -s^\top Q s - P_1 |s_1| - P_2 |s_2| \tag{4.63}$$

For $\dot{V}$ to be negative semi-definite, we choose Q_i and P_i such that $Q_i P_i \geq 0$. Then, given that $V > 0$ and that $\dot{V} \leq 0$, the control law is stable in the Lyapunov sense. Finally, the single-wheel velocity commands for the mobile robot are practically given by

$$\begin{cases} \Omega_r = \frac{v_c + b\omega_c}{r} \\ \Omega_r = \frac{v_c - b\omega_c}{r}, \end{cases} \tag{4.64}$$

where, following the conventions in Fig. 4.11, r is radius of the driving wheels, b is half the distance between the driving wheels, v_c is the computed control velocity, and ω_c the computed control angular velocity (see Eqs. (4.59) and (4.60)). These values are subsequently sent to the inner loop of the closed-loop system in Fig. 4.13, more precisely to the PID controllers (separately treated in the previous section and Fig. 4.14) where the encoder revolutions N_r and N_l are available from odometric computations.

4.2.3.2 Experimental Results

To validate the described *Antifragile* controller, we consider the trajectory tracking problem shown in Fig. 4.17a. The controller is also compared against three baseline approaches: (i) a *Robust* controller based on the approach in [38]; (ii) an *Adaptive* receding horizon controller based on the model predictive controller presented in [77]; and (iii) a *Resilient* controller based on the fuzzy logic controller proposed

4.2 Antifragile Robotic Systems

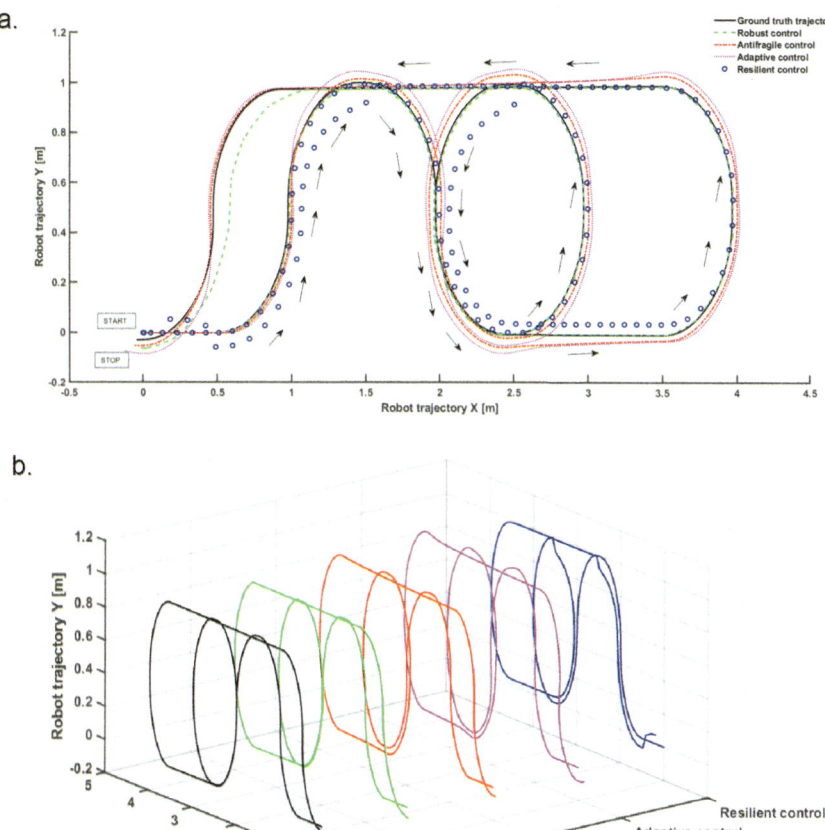

Fig. 4.17 Trajectory tracking control: **a** Experimental trajectory description and individual controllers operation. **b** Comparison of the control approaches and their performance in tracking the desired trajectory

in [78]. The reader interested in implementation details is referred to the open source codebase available on GitHub.[1]

Analyzing the results in Fig. 4.17, we can observe that each control strategy overshoots when tracking the reference trajectory, but as we will see in the experiments this is a compensation mechanism for the curvature of the trajectory with more capacity to handle uncertainties (i.e., Antifragile and Adaptive in panel a)). Additionally, we see undershooting behavior in control approaches, which are tracking almost

---

[1] Codebase available at https://gitlab.com/akii-microlab/antifragile-robot-control/.

perfectly the low curvature regions (i.e., Robust and Resilient in panel a) but then have no capacity when the curvature increases (e.g., see the end of trajectory and the inner cycle).

4.2.4 Conclusion

The modelling and handling of uncertainty in closed-loop robot control tasks remains a topic of considerable debate and fruitful research. In the context of control theory providing its most powerful tools and robotics offering its more pragmatic deployments, emerging approaches must overcome well-established "recipes". The concept of antifragile control represents a novel approach to control theory. It is based on the idea of capturing the distinctive characteristics of a system's response to control inputs. The aim is to identify control signals that can steer the system towards regions of the solution space where it is robust to uncertainty and volatility. In these regions, the system can not only withstand but also benefit from these conditions and anticipate future uncertain events. This is the fundamental rationale behind the antifragile control approach.

The current study is an exploratory one, along with the previous instantiations of Antifragile control in [15, 58], and the objective is to prompt the community to adopt and leverage a novel control system design where crucial design information is embedded in metrics that describe the shape of the system response to uncertainty. In the present iteration of the Antifragile control system for mobile robot trajectory tracking, we have merely begun to explore the potential of this framework. The experiments with parameterizable faults have enabled us to validate the framework and the controller design for a relatively simple task and dynamics. We are interested in developing a coherent thesis and framework based on Antifragile control principles and facilitating the introduction of antifragility in technical systems.

4.3 Antifragile Stability Analysis of Nonlinear Control Systems

One of the most critical issues in control is determining the stability of a system. Since the 1960s, Lyapunov-based methods have been developed to determine the stability of linear and nonlinear systems. However, when the system is nonlinear, time-dependent and uncertain, in a set-membership context, stability analysis is challenging, and no reliable methods have been developed. This section proposes an original antifragile set-membership-based approach for establishing the stability of non-linear, uncertain, time-dependent systems. Two new concepts, Antifragile Stability, AF-Stability (which is the stability of nonlinear time-dependent trajectories) and Antifragile Tubes, AF-Tubes (which are invariant stability regions for time-dependent systems) are introduced in the context of antifragility and illustrated for an autonomous robotic sailboat. Then, AF-Stability is used to formulate and prove the safety for a set of antifragile tubes. This result is then used to analyze the safety of a squad of uncertain, robotic sailboats moving in their environment, i.e., no collision among the robots. Examples second the conceptual work to offer the reader "recipes" to design antifragile stable feedback control loops.

4.3.1 Preliminaries

The concept of stability was initially formulated in the late 19th century by the Russian mathematician Aleksandr Mikhailovich Lyapunov [79, 80]. This came to the attention of the control engineering community in the 1960s, and since then, rigorous stability methods have been developed for both linear [81, 82] and nonlinear [83, 84] systems. In particular, in the last 20 years, researchers have been developing methods for determining the stability of time-invariant uncertain systems, using set-membership techniques [85, 86].

A time-dependent, nonlinear, dynamic system can be described by a state equation $\dot{x} = f(x, t)$ or, if the system is uncertain, by a time-dependent deferential inclusion [87], $\dot{x} \in F(x, t)$, where x is the state vector and t represents time. Analyzing the stability properties of this system (or of the differential inclusion) is an important yet difficult problem. For some particular properties and for time-invariant systems $\dot{x} = f(x)$ or time-invariant differential inclusions $\dot{x} \in F(x)$ (i.e., for F do not depend on t). It has been shown that the V-Stability approach [82, 87] (which is derived from Lyapunov theory for stability analysis of nonlinear systems) combined with interval analysis [88] can solve the stability problem. The main idea is to prove that the solution of $\dot{x} = f(x)$ (or the solutions of $\dot{x} \in F(x)$) will always stay inside a bubble (i.e., inside an invariant set).

Our goals in this section are twofold. Firstly, to show that proving V-Stability can be transformed into an antifragility problem by proving the inconsistency for a set of inequalities. Secondly, we want to introduce two new concepts, Antifragile Tubes and Antifragile Stability, to deal with stability problems for time-dependent systems or time-dependent deferential inclusions.

To the best of our knowledge, when a system is nonlinear, time-dependent and uncertain, in a set-membership context, no reliable methods are available for determining its stability. To prove that $\dot{x} \in F(x, t)$ is AF-Stable amounts to proving that all solutions of $\dot{x} \in F(x, t)$ will always stay inside a time-varying bubble (i.e., an antifragile tube). It will be shown that if an antifragile tube candidate can be calculated for a time-dependent system (or differential inclusion), then the system (or the differential inclusion) is AF-Stable. Moreover, using these concepts, a new method is proposed to prove the safety of a set of antifragile tubes.

The section is organized as follows. We start showing how the V-Stability problem can be transformed into an antifragility problem by proving the inconsistency of a set of inequalities. It will also be shown that interval analysis and contractors

[82] can efficiently solve this problem in polynomial time. Afterwards, two new concepts to analyze the stability problem for time-dependent systems or deferential inclusions, antifragile tubes and AF-Stability, will be introduced. Based on the same idea proposed for V-Stability, it will be shown that proving that a tube is an antifragile tube candidate can be transformed into proving the inconsistency of a set of inequalities. These new concepts will then be used to establish the safety of a set of antifragile tubes. To support the theoretical development, two case studies are described to illustrate the application of both AF-Stability for a sailboat robot, adapted from [89] and the safety of a squad of sailboat robots.

4.4 Antifragile Stability

The idea of Antifragile Stability is derived from Lyapunov stability theory and influenced by the book of Aubin and Frankowska [87] and by the V-Stability concept proposed by Jaulin in his seminal paper [82]. Before introducing the Antifragile Stability concept, recall the Lyapunov stability theorem for time-invariant systems, described by Eq. (4.65). Let $x = 0$ be an equilibrium point of $\dot{x} = f(x)$, and let V be a continuously differentiable function defined in a domain $D \subset R^n$, $V : D \to R$ such that:

$$V(0) = 0,$$
$$\dot{V}(x) > 0 \in D - \{0\}, \quad (4.65)$$
$$\dot{V}(x) \leq 0 \in D - \{0\},$$

then $x = 0$ is stable. Lyapunov stability theory requires that $V(x)$ is always positive, except at the equilibrium point. V-Stability extends Lyapunov stability to the case where $V(x)$ is not always positive, i.e., it can also be negative for some x.

4.4.1 Antifragile Stability for Time-Invariant Dynamic Systems

Definition 4.1 Consider a differentiable function $V : R^n \to R$. The system $\dot{x} = f(x)$ is said to be Antifragile Stable (AF-Stable) if there exist $\varepsilon > 0$ such that:

$$\text{If } V(x) \geq 0 \Rightarrow \dot{V}(x) \leq -\varepsilon < 0$$
$$\text{where } \dot{V}(x) = \frac{\partial V}{\partial x}(x) \cdot f(x). \quad (4.66)$$

The function $V(x)$ appears similar to a Lyapunov function candidate in Eq. (4.66), but, in reality, it is not exactly the same. The concept of AF-Stability is illustrated in Fig. 4.18, where the arrows represent the vector field of the dynamical system.

4.4 Antifragile Stability

Fig. 4.18 V-Stability for $V(x) \in [0, \infty]$

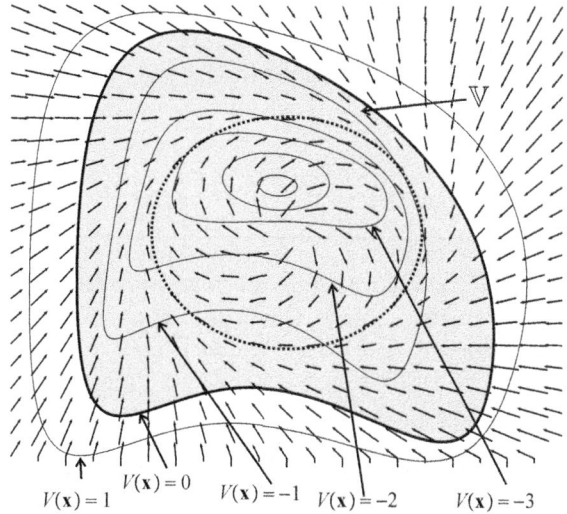

When $V(x) \geq 0$, then all system trajectories must be attracted by the grey region (bubble) limited by the level curve $V(x) = 0$, which is similar to Lyapunov stability because $\dot{V}(x) \leq -\varepsilon < 0$. As in the case of V-Stable systems [86], the difference arises when the state lies in the gray bubble where $V(x)$ becomes negative. Then, the system trajectory can follow any shape, for instance, a limit cycle, as illustrated in the figure. The AF-Stability definition states that when the scalar differentiable function $V(x)$ is positive, then it will strictly decrease along the level curves. In this case, the trajectory will be attracted by the grey region, which is an invariant set. Once inside the attraction zone, the trajectory will stay inside forever and can follow any behavior. Consequently, the notion of V-Stability is weaker than the stability in the sense of Lyapunov [90].

4.4.2 Consequence of the Lyapunov Theorem

The stability problem of a dynamic system can be represented as a set of inequalities. This is very important because the set of inequalities can be easily solved using numerical methods. This chapter proposes a similar idea by showing that AF-Stability can be determined from the inconsistency of a set of inequalities, as is proposed in Theorem 4.1.

Theorem 4.1 *Consider a nonlinear time-invariant system $\dot{x} = f(x)$. If the set of constraints:*

$$\begin{cases} \frac{\partial V}{\partial x}(x) \cdot f(x) \geq 0 \\ V(x) \geq 0 \end{cases}$$

is inconsistent $\Leftrightarrow \dot{x} = f(x)$ is AF-Stable.

Proof Assume that $\dot{x} = f(x)$ is V-Stable. From Eq. (4.66), $V(x) \geq 0 \Rightarrow \dot{V}(x) < 0$, where V is a scalar differentiable function. As the state vector x depends on time, this implies that the function V will also depend on time. Differentiating the function V with respect to time gives $\frac{\partial V(x)}{\partial t} = \frac{\partial V}{\partial x}(x)\dot{x} = \frac{\partial V}{\partial x}(x) \cdot f(x)$. AF-Stability becomes $V(x) \geq 0 \Rightarrow \frac{\partial V}{\partial x}(x) \cdot f(x) < 0$, which is true for all x. Using the logical relationship $(A \Rightarrow B)$ [27FA°] $(B \vee \neg A)$, where A is $V(x) \geq 0$ and B is $\frac{\partial V}{\partial x}(x) \cdot f(x) < 0$, the implication is equivalent with:

$$V(x) \geq 0 \Rightarrow \frac{\partial V}{\partial x}(x) \cdot f(x) < 0 \Leftrightarrow \frac{\partial V}{\partial x}(x) \cdot f(x) < 0 \vee V(x) < 0$$

Using De Morgan's law $\overline{A \vee B} = \bar{A} \wedge \bar{B}$, where $\vee$ and $\wedge$ represent the logical OR and AND operators, respectively, the AF-Stability problem can be represented as:

$$\neg \exists x, \left(\frac{\partial V}{\partial x}(x) \cdot f(x) \geq 0 \wedge V(x) \geq 0 \right)$$

which proves the Theorem. □

A simple example of the determination of AF-Stability for the time-invariant linear system $\dot{x} = -x + 1$ is now described. Consider $V(x) = x^2 - r^2$ as an AF function candidate. It can be easily observed that $V(x)$ is not always positive and, using Theorem 4.1, it will be shown that the system is AF-Stable for $r = 1$. AF-Stability is equivalent to finding r such that $V(x)$ always converges to a negative value, i.e., $V(x) = x^2 - r^2 \leq 0$. The rate of change of the scalar field V along the flow of the vector field $\dot{x}$ is $\frac{\partial V}{\partial x}(x) = 2x\dot{x} = 2x(-x + 1)$. The AF-Stability problem becomes

$$\begin{cases} 2x(-x+1) < 0 \\ x^2 - r^2 \geq 0 \end{cases}$$

which based on Theorem 4.1 becomes

$$\begin{cases} 2x(-x+1) \geq 0 \\ x^2 - r^2 \geq 0 \end{cases}$$

which is inconsistent for $r \geq 1$ and therefore the system is AF-Stable for $r \geq 1$. The bubble $x^2 - r^2$ (with $r = 1$) becomes negative for $x < 1$ which is an attractive bubble for the system. Inside this bubble, it is easy to see that $\dot{V}$ is positive for $x \in [0, 1]$

and negative for $x \in [-1, 0]$, and as it was previously stated, the system behavior is unimportant inside the bubble. Outside the bubble, $\dot{V}$ is only negative.

4.4.3 A F-Stability for Time-Invariant Differential Inclusions

As uncertainties are present in real systems, it is more realistic to represent the dynamical systems as differential inclusions [91]. This notion makes it possible to develop numerical algorithms to rigorously study the stability [92, 93]. In this section, the concept of differential inclusion is introduced, and then it is shown that Theorem 4.1 can be extended for uncertain systems. Furthermore, it is also shown how AF-Stability can be efficiently determined using interval analysis and contractors [82]. Differential inclusions are a generalization of the concept of the state equation and are used to represent uncertain dynamic systems in a set-membership framework. A differential inclusion can be defined by the following inclusion $\dot{x} \in F(x)$, where F is a multi-valued function from R^n to R^n. A multi-valued function $F: R^n \to R^n$ associates each $x \in R^n$ with a subset $F(x)$ of R^n [87, 94].

Definition 4.2 Consider a differentiable function $V(x) : R^n \to R$. The differential inclusion $\dot{x} \in F(x)$ is AF-Stable if there exist $\varepsilon > 0$ such that

$$\text{If } V(x) \geq 0 \Rightarrow \dot{V}(x) \leq -\varepsilon < 0$$

$$\text{where } \dot{V}(x) = \frac{\partial V}{\partial x}(x) \cdot a, \text{ and } a \in F(x)$$

Theorem 4.2 *Consider an uncertain system represented by a time-invariant differential inclusion $\dot{x} \in F(x)$. If the system of constraints*

$$\begin{cases} \frac{\partial V}{\partial x}(x) \cdot a \geq 0 \\ a \in F(x) \\ V(x) \geq 0 \end{cases}$$

is inconsistent $\Leftrightarrow \dot{x} \in F(x)$ is AF-Stable.

Proof Assume that $\dot{x} \in F(x)$ is AF-Stable. From Definition 4.2, $V(x) \geq 0 \Rightarrow \dot{V}(x) < 0$, where V is a scalar function. Differentiating V with respect to time: $\frac{\partial V(x)}{\partial t} = \frac{\partial V}{\partial x}(x) \cdot a$ where $a \in F(x)$, AF-Stability becomes: $V(x) \geq 0 \Rightarrow \frac{\partial V}{\partial x}(x) \cdot a < 0 \ \forall a \in F(x)$. Using the logical rule $(A \Rightarrow B)(B \vee \neg A)$ the following relationship exists:

$$\forall x \in [x], \forall a \in F(x), \left(\frac{\partial V}{\partial x}(x) \cdot a < 0 \bigvee V(x) < 0 \right)$$

and applying De Morgan's laws:

$$\neg \exists x \in [x], \neg \exists a \in F(x), \left(\frac{\partial V}{\partial x}(x) \cdot a \geq 0 \wedge V(x) \geq 0\right)$$

which proves the theorem. □

Consequence Theorems 4.1 and 4.2 show how the AF-Stability problem can be transformed into a set of inequalities. Using interval analysis and contractors, it is possible to prove that a time-invariant differential inclusion is AF-Stable by efficiently establishing the inconsistency of a set of inequalities.

In the following, the efficiency of using contractors to prove the inconsistency of a set of inequalities is illustrated. The computational complexity is polynomial-time with respect to the number of variables. Consider the following set of inequalities:

$$\begin{cases} f(x) = x^2 - 1 \geq 0 \\ f(x) = \frac{-1}{x} \geq 0 \\ f(x) = 2x + 1 \geq 0 \end{cases}$$

where x is defined on the domain $[-\infty, \infty]$. Using interval propagation [82], it is required to show that the system of given inequalities is inconsistent, i.e., has no solution. A contractor-based method is a very powerful tool to prove the inconsistency of a set of inequalities. A contractor C is an operator that uses a constraint satisfaction problem, as shown in [82], to compute the subset $[x']$ such that the solution set S remains unchanged, i.e., $S \subseteq [x'] \subseteq [x]$. An interval of R is a closed-connected subset of R. A box $[x]$ of R^n is the Cartesian product of n intervals. The set of all boxes of R^n is denoted by IR^n. A contractor C is an operator $IR^n \to IR^n$ such that satisfies the contractance and correctness properties as shown in Eq. (4.67).

$$\begin{aligned} \forall [x] \in IR^n, C([x]) \subset [x] \quad &\text{contractance} \\ C([x]) \cap S = [x] \cap S \quad &\text{correctness} \end{aligned} \quad (4.67)$$

The inequality set in the example can be transformed into a constraint satisfaction problem by introducing a slack variable y. Using this slack variable, the set of inequalities (11) is transformed into the following set of constraints

$$\begin{cases} x^2 - 1 - y = 0 \\ \frac{-1}{x} - y = 0 \\ 2x + 1 - y = 0 \end{cases}$$, where the domain for the slack variable y is $[0, \infty]$. Using interval propagation, it is necessary to prove that the set of equations

$$\begin{cases} (C_1): y = x^2 - 1 \\ (C_2): y = \frac{-1}{x} \\ (C_3): y = 2x + 1, \end{cases}$$ has no solution. When several constraints are involved, the contractions are performed sequentially until no further contraction can be made. The domains are contracted in the following order: C_1, C_2, C_3, C_1 until empty intervals for x and y are obtained. The resulting interval computation is as follows:

$(C_1) :\Rightarrow y \in [0, \infty] \cap ([-\infty, \infty]^2 - 1) \Rightarrow y \in [0, \infty] \cap [-1, \infty] = [0, \infty]$

$(C_2) :\Rightarrow x = \dfrac{-1}{y} \Rightarrow x \in \dfrac{-1}{[0, \infty]} = [-\infty, 0]$

$(C_3) :\Rightarrow y \in [0.\infty] \cap (2 \cdot [-\infty, 0] + 1) = [0, 1]$ and $x \in [-\infty, 0] \cap \left(\dfrac{[0,1]}{2} - \dfrac{1}{2}\right) = \left[\dfrac{-1}{2}, 0\right]$

$(C_1) :\Rightarrow y \in [0, 1] \cap \left(\left[\dfrac{-1}{2}, 0\right]^2 - 1\right) = [0, 1] \cap \left[-1, -\dfrac{3}{4}\right] = \emptyset$

An important remark is that the interval propagation method converges to a box that encloses all solutions (if any exist), but the box is not necessarily minimal. The box is said to be locally consistent because it is consistent with all constraints taken independently. The smallest box that encloses all solutions is said to be globally consistent. The problem of computing this smallest box is NP-hard and can be solved using bisection methods only for problems involving a small number of variables [82]. This is not the case for the methods proposed in this section, where the aim is to establish inconsistency.

4.5 Stability Analysis for Time-Dependent Systems and Time-Dependent Differential Inclusions

In this section, the concept of AF-Stability is extended to systems where f and F both depend on time. In Sect. 4.5.1, the notion of antifragile tubes will be introduced and used in Sect. 4.5.2 to define AF-Stability for time-dependent systems and differential inclusions. In Sect. 4.5.3, the notion of safety for a set of antifragile tubes is introduced.

4.5.1 Antifragile Tubes

Consider a non-autonomous, nonlinear, time-dependent system described by the state equation $\dot{x} = f(x, t)$ with a target condition. This can be described by:

$$\begin{cases} S_f : \dot{x} = f(x, t) \text{ (evolution equation)} \\ g(x, t) \leq 0 \text{ (target condition)} \end{cases} \quad (4.68)$$

where $x \in R^n$ is the state vector, $f : R^n \to R^n$ is the evolution function and $g : R^n \times R \to R^m$ is the target function. At every time instant, the target condition Eq. (4.68)

Fig. 4.19 Representing a tube in state space as a time moving bubble

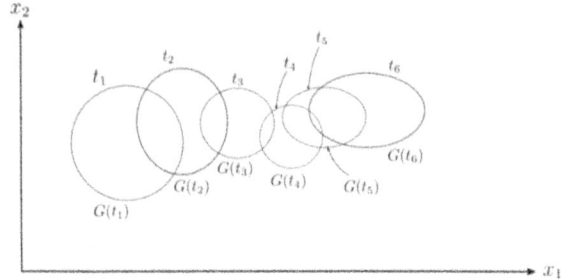

Fig. 4.20 A tube in the time domain

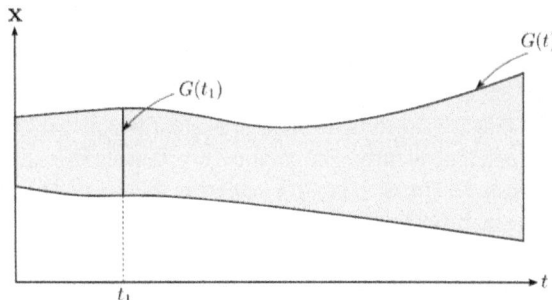

is a bubble similar to the grey region in Fig. 4.18. Since the target condition is time-dependent, the bubble moves in time and will generate an antifragile tube, represented by inequalities, $G(t) = \{x, g(x, t) \leq 0\}$ which associates with each value of $t \in R$ a subset of R^n. A tube (or interval of trajectories), according to [95, 96], is a set-membership vision of a random signal. The extension of classical operations on trajectories (such as sums, multiplications, image by a function) to tubes is presented in [97]. In Figs. 4.19 and 4.20, two possible representations for a tube are shown. Figure 4.19 shows a bubble that moves with respect to time, which is represented at six different time instants in state space. Another possible representation for a tube is in the time domain as the grey tube in Fig. 4.20. $G(t_1)$ from Fig. 4.20 is the bubble at time instant t_1, as illustrated in Fig. 4.19.

Definition 4.3 A trajectory is stable if it satisfies conditions in Eq. (4.68). The system $\dot{x} = f(x)$ is said to be stable if for all $x(0) \in G(0)$, the corresponding trajectory is stable.

Definition 4.4 A target tube is said to be an antifragile tube for S_f if

$$x(t) \in G(t), \forall t_1 > 0 \Rightarrow x(t + t_1) \in G(t + t_1)$$

Figure 4.21 illustrates some feasible trajectories and a tube $G(t)$ (in grey). All trajectories are consistent with the assumption that $G(t)$ is an antifragile tube, except for the trajectory which is represented by a dotted curve at the bottom, which was able to escape from the tube.

4.5 Stability Analysis for Time-Dependent Systems and Time-Dependent ...

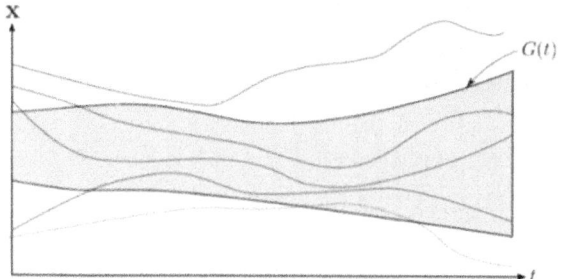

Fig. 4.21 A target tube (colored grey) and several trajectories for different initial conditions

Taking into account that a tube can be represented by inequalities, the following theorem shows that the problem of proving that $G(t)$ is an antifragile tube can be reformulated as proving that a set of inequalities has no solution.

Theorem 4.3 *Consider a tube* $G(t) = \{x, g(x, t) \leq 0\}$ *where* $g : R^n \times R \to R^m$. *If the cross out condition in Eq.* (4.68)

$$\begin{cases} (i) \dfrac{\partial g_i}{\partial x}(x, t) \cdot f(x) + \dfrac{\partial g_i}{\partial t}(x, t) \geq 0 \\ (ii) g_i(x, t) = 0 \\ (iii) g(x, t) \leq 0 \end{cases} \quad (4.69)$$

is inconsistent for all x, *all* $t > 0$, *and all* $i \in \{1, \ldots, m\}$ *then* $G(t)$ *is an antifragile tube for the time-dependent system* $\dot{x} = f(x, t)$.

Proof The proof for Theorem 4.69 is by contradiction. Assume that $G(t)$ is not a capture tube. It means that there exists one trajectory which leaves the tube $G(t)$, i.e., which crosses the i-th boundary $g_i(x, t) = 0$ from inside to outside. This means that there exists a time-space pair (s, t_s) on the boundary of $G(t)$, such that (ii) and (iii) are satisfied, and $\dot{g}_i(x, t)x = s$ (otherwise the trajectory cannot leave the tube). Thus, if $G(t)$ is not an antifragile tube, there exists one i such that the system of constraints

$$\begin{cases} \dfrac{\partial g_i}{\partial x}(x, t_1) \cdot f(x, t_1) x = s \\ t = t_s \end{cases} \quad (4.70)$$

is consistent, which is in contradiction with the assumption of the Theorem 4.69. □

An illustrative example is given to show how Theorem 4.69 can be used to prove that a tube is an antifragile tube for a simple time-dependent function. Consider a time-dependent, linear, first order system

$$\dot{x} = -x + t^2 \quad (4.71)$$

Using Theorem 4.3, one prove that the tube $g(x, t) = (x - t^2)^2 - 1 \leq 0$ is an antifragile tube for the system in Eq. (4.71). This will be done by proving the inconsistency of the following set of inequalities:

$$\begin{cases} (i) \frac{\partial g}{\partial x}(x,t) \cdot f(x,t) + \frac{\partial g}{\partial t}(x,t) \geq 0 \Leftrightarrow 2(x - t^2)(-x + t^2) + 2(x - t^2)(-2t) \geq 0 \\ (ii) g(x,t) = 0 \Leftrightarrow (x - t^2)^2 - 1 = 0 \\ (iii) g(x,t) \leq 0 \Leftrightarrow (x - t^2)^2 - 1 \leq 0 \end{cases}$$

The previous set of inequalities becomes

$$\begin{cases} (i) (x - t^2)^2 + 2(x - t^2)t \leq 0 \\ (ii) x - t^2 = 1 \end{cases}$$

$$\Leftrightarrow$$

$$\begin{cases} (i) 1 + 2t \leq 0 \\ (ii) x - t^2 = 1 \end{cases}$$

which is inconsistent, therefore $g(x, t)$ is an antifragile tube for the time-dependent system (4.71). This means that if the system is initialized inside the antifragile tube at $t = 0$, it will remain inside the antifragile tube for all $t > 0$, i.e., the cross-out condition is not satisfied. For a complex, nonlinear, differential inclusion, it is not easy to prove stability by solving the set of inequalities. In this case, a contractor-based method will be used, as it was shown in previous examples.

If the system is both time-dependent and uncertain, in a set-membership context, Theorem 4.3 can be extended for a time-dependent differential inclusion

$$S_F : \dot{x} \in F(x, t) \tag{4.72}$$

as in Theorem 4.4.

Theorem 4.4 *Consider a tube $G(t) = \{x, g(x, t) \leq 0\}$ where $g : R^n \times R \to R^m$. If the cross out condition*

$$\begin{cases} (i) \frac{\partial g_i}{\partial x}(x,t) \cdot a + \frac{\partial g_i}{\partial t}(x,t) \geq 0 \\ (ii) a \in F(x) \\ (iii) g_i(x,t) = 0 \\ (iv) g(x,t) \leq 0 \end{cases} \tag{4.73}$$

is inconsistent for all x, all a, all $t > 0$, and all $i \in \{1, \ldots, m\}$ then $G(t) = \{x, g(x, t) \leq 0\}$ is an antifragile tube for the time dependent differential inclusion $\dot{x} \in F(x, t)$.

4.5 Stability Analysis for Time-Dependent Systems and Time-Dependent ...

Proof The proof of Theorem 4.4 is by contradiction. Assume that $G(t)$ is not an antifragile tube. Hence, there exists one trajectory $a \in F(x)$ which leaves the tube $G(t)$, i.e., which crosses the ith boundary $g_i(x, t) = 0$ from inside to outside. This means that there exists a time-space pair (s, t_s) on the boundary of $G(t)$, such that (iii) and (iv) are satisfied, and $\dot{g}_i(x, t)x = s$ (otherwise the trajectory cannot leave the tube). Thus, $G(t)$ is not an antifragile tube, there exists one i such that the system of constraints

$$\begin{cases} \frac{\partial g_i}{\partial x}(x, t_1) \cdot ax = s \\ t = t_s \end{cases}$$

is consistent, which is in contradiction with the assumption of the theorem. □

As a consequence, an antifragile tube is a special type of tube where, for a dynamic system, once the state is inside the tube, it will stay inside forever. From Theorems 4.3 and 4.4, checking that a tube, defined by a set of inequalities, is an antifragile tube amounts to checking the inconsistency of a set of inequalities. This can easily be performed using contractor-based methods, as was shown in the last example.

In the last example, we consider the antifragile tube in the time domain. Consider again Fig. 4.21, where it is assumed that the grey tube corresponds to $G(t) = \{x, g_1(x, t) \leq 0\}$. The dotted trajectory leaves the tube at a time-space point (s, t_s), such that $g_1(s, t) = 0$ and $\dot{g}_1(s, t) > 0$. In this case, the tube is not an antifragile tube since there exists a trajectory which leaves the tube, i.e., the tube is not an invariant set.

4.5.2 *AF-Stability for Time-Dependent Systems S_f and S_F*

To generalize the framework, we provide in the following a set of definitions and theorems that will support the reader in grasping the fundamentals revamped in the examples at the end of the chapter.

Definition 4.5 Consider a tube $G(t) = \{x, g(x, t) \leq 0\}$ where $g : R^n \times R \to R^m$. S_f is said to be AF-Stable if:

$$\begin{cases} g_i(x, t) = 0 \\ g(x, t) \leq 0 \end{cases} \Rightarrow \dot{g}(x, t) < 0$$

where $\dot{g}(x, t) = \frac{\partial g}{\partial x}(x, t) \cdot f(x, t_1) + \frac{\partial g}{\partial t}(x, t)$

Theorem 4.5 *If a target tube $G(t) = \{x, g(x, t) \leq 0\}$ for a time-dependent nonlinear system (4.68) is an antifragile tube, then the system (4.68) is AF-Stable.*

Proof Assume that $\dot{x} = f(x, t)$ is AF-Stable. Definition 4.5 gives

$$\begin{cases} g_i(x, t) = 0 \\ g(x, t) \leq 0 \end{cases} \Rightarrow \dot{g}(x, t) < 0$$

where $g(x,t)$ is a differentiable target tube and $\dot{g}(x,t) = \frac{\partial g}{\partial x}(x,t) \cdot f(x,t_1) + \frac{\partial g}{\partial t}(x,t)$. Now, using the logical rule $(A \Rightarrow B)(B \vee \neg A)$ where A is given by

$$\begin{cases} g_i(x,t) = 0 \\ g(x,t) \le 0 \end{cases}$$

B is $\dot{g}(x,t) < 0$, and applying De Morgan's laws for

$$\begin{cases} g_i(x,t) = 0 \\ g(x,t) \le 0 \end{cases}$$

then the implication is equivalent with

$$\dot{g}(x,t) < 0 \bigvee g_i(x,t) \ne 0 \bigvee g(x,t) > 0$$

Applying again De Morgan's laws for the last implication, one gets:

$$\neg \exists x, \left(\dot{g}(x,t) \ge 0 \bigwedge g_i(x,t) = 0 \wedge g(x,t) \le 0 \right)$$

or, equivalently, the system of constraints

$$\begin{cases} (i)\ \frac{\partial g_i}{\partial x}(x,t) \cdot f(x) + \frac{\partial g_i}{\partial t}(x,t) \ge 0 \\ (ii)\ g_i(x,t) = 0 \\ (iii)\ g(x,t) \le 0 \end{cases}$$

is inconsistent. The system of inequalities in Definition 4.5 represents the cross out condition. Since this is inconsistent with Theorem 4.3, $G(t) = \{x, g(x,t) \le 0\}$ is an antifragile tube for (4.68), which proves the theorem. $\square$

Definition 4.6 Consider a tube $G(t) = \{x, g(x,t) \le 0\}$ where $g : R^n \times R \to R^m$. The time dependent differential inclusion of Eq. (4.72) is said to be AF-Stable if:

$$\begin{cases} g_i(x,t) = 0 \\ g(x,t) \le 0 \end{cases} \Rightarrow \dot{g}(x,t) < 0$$

where $\dot{g}(x,t) = \frac{\partial g}{\partial x}(x,t) \cdot a + \frac{\partial g}{\partial t}(x,t)$ and $a \in F(x)$.

Theorem 4.6 *If a target tube $G(t) = \{x, g(x,t) \le 0\}$ for the time-dependent differential inclusion $\dot{x} \in F(x,t)$ is an antifragile tube, then the differential inclusion $\dot{x} \in F(x,t)$ is AF-Stable.*

Proof Assume that $\dot{x} \in F(x,t)$ is AF-Stable. Definition 4.6 gives

$$\begin{cases} g_i(x,t) = 0 \\ g(x,t) \le 0 \end{cases} \Rightarrow \dot{g}(x,t) < 0$$

4.5 Stability Analysis for Time-Dependent Systems and Time-Dependent ...

where $g(x, t)$ is a differentiable target tube and $\dot{g}(x, t) = \frac{\partial g}{\partial x}(x, t) \cdot a + \frac{\partial g}{\partial t}(x, t)$ and $a \in F(x)$. Now, using the logical rule $(A \Rightarrow B)(B \vee \neg A)$ where A is

$$\begin{cases} g_i(x, t) = 0 \\ g(x, t) \leq 0 \end{cases}$$

B is $\dot{g}(x, t) < 0$ and applying De Morgan's laws for

$$\begin{cases} g_i(x, t) = 0 \\ g(x, t) \leq 0 \end{cases}$$

then the implication is equivalent with

$$\forall x \in [x], \forall a \in F(x, t), (\dot{g}(x, t) < 0 \bigvee g_i(x, t) \neq 0 \bigvee g(x, t) > 0)$$

Applying De Morgan's laws for the last implication relation one gets:

$$\neg \exists x \in [x], \neg \exists a \in F(x), (\dot{g}(x, t) \geq 0) \bigwedge g_i(x, t) = 0 \bigwedge g(x, t) \leq 0)$$

or equivalently, the system of constraints

$$\begin{cases} (i) \dfrac{\partial g_i}{\partial x}(x, t) \cdot a + \dfrac{\partial g_i}{\partial t}(x, t) \geq 0 \\ (ii) a \in F(x) \\ (iii) g_i(x, t) = 0 \\ (iv) g(x, t) \leq 0 \end{cases}$$

is inconsistent. The system of inequalities above is the cross out condition and, since it is inconsistent with Theorem 4.4, $G(t) = \{x, g(x, t) \leq 0\}$ is an antifragile tube for differential inclusion $\dot{x} \in F(x, t)$ which proves the theorem. □

Finally, as a consequence, an antifragile tube is a capture set which depends on time. From Theorems 4.5 and 4.6, it can be concluded that when there exists an antifragile tube for S_f or for S_F, then the system or the differential inclusion is AF-Stable in the sense of Definitions 4.5 or 4.6. In other words, if S_f (or S_F) is AF-Stable then the trajectories of S_f (or S_F) will be captured by the antifragile tube $G(t) = \{x, g(x, t) \leq 0\}$, i.e., will stay forever inside the antifragile tube. To better support the formalism, Fig. 4.22 illustrates the notions of antifragile tube and AF-Stability for a two-dimensional target tube. The antifragile tube is represented in the figure, at a particular time instant t, by a two dimensional capture bubble $G(t)$ (grey bubble), as the intersection of the two bubbles, $g_1(x, t) \leq 0$ and $g_2(x, t) \leq 0$, at the same time instant. $G(t)$ is a capture bubble for S_f or S_F, (i.e., S_f or S_F are AF-Stable) if for every state inside $G(t)$, it will remain there forever. The system

Fig. 4.22 The cross out condition for a two dimensional antifragile tube

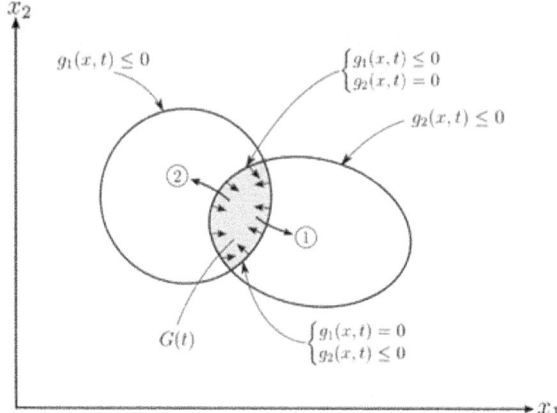

trajectories can cross out the boundaries of $G(t)$ only through two regions, i.e., either through the frontier

$$\begin{cases} g_1(x,t) \leq 0 \\ g_2(x,t) = 0 \end{cases}$$

or through the frontier

$$\begin{cases} g_1(x,t) = 0 \\ g_2(x,t) \leq 0 \end{cases}$$

(see Fig. 4.22). Theorems 4.3 and 4.4 answer the following question: Is it possible to leave the grey region $G(t)$? This only happens if there exists a time-space pair (s, t_s) on the boundary of $G(t)$ such that $\dot{g}_i(x,t) \geq 0$ (otherwise the trajectory cannot leave the tube), as in the case of the trajectories represented in Fig. 4.22 with Trajectory 1 and Trajectory 2, i.e., $\dot{g}_1(x_1,t) > 0$ or $\dot{g}_2(x_2,t) > 0$, respectively. In this case, the cross-out condition from Theorems 4.3 or 4.4 becomes consistent and $G(t)$ is not an antifragile bubble, i.e., S_f or S_F are not AF Stable. If the cross-out condition from Theorems 4.3 or 4.4 is inconsistent, this means that the Trajectory 1 and Trajectory 2 are not feasible and the only feasible trajectories are the ones generated by the vector field represented by arrows oriented towards the inside of the grey region, i.e., $\dot{g}_1(x,t) < 0$ and $\dot{g}_2(x,t) < 0$. In this case, the trajectories are always inside the bubble, which implies that $G(t)$ is an antifragile bubble S_f (or S_F) is AF-Stable and. This reasoning is valid for any time $t > 0$.

4.5.3 Safe Antifragile Tubes

In the following, we extend the formalism with the safe antifragile tubes in the general systems case.

Definition 4.7 For a set of n AF-Stable differential inclusions $\dot{x}^a \in F^a(x)$, the afferent capture tubes $G^a(t) = \{x, g^a(x,t) \leq 0\}$, $a \in \{1,\ldots,n\}$ are safe if any trajectory in $G^a(t)$ does not intersect any other trajectory in $G^b(t)$, $a, b \in \{1,\ldots,n\}$, $a \neq b$.

Based on the results presented in Sects. 4.5.1 and 4.5.2, the following theorem shows that proving that a set of antifragile tubes are safe can be transformed in proving the inconsistency of a set of inequalities.

Theorem 4.7 *Consider a set of p AF-Stable nonlinear time-dependent systems $S_f^1, \ldots, S_f^p$ (or time dependent differential inclusions $S_F^1, \ldots, S_F^p$) and the associated antifragile tubes: $G^1(t), \ldots, G^p(t)$. If for all $t \geq 0$, for all $a, b \in \{1, \ldots, p\}$, $a \neq b$, and for all x^a, the system of constraints*

$$\begin{cases} (i) \ g^a(x^a, t) \leq 0 \\ (ii) \ g^b(x^a, t) < 0 \end{cases} \quad (4.74)$$

is inconsistent, then the antifragile tubes $G^1(t), \ldots, G^p(t)$ are safe.

Proof The proof is by contradiction. Assume that the antifragile tubes are not safe. From the theorem, each nonlinear time-dependent system $S_f^1, \ldots, S_f^p$ (or time dependent differential inclusions $S_F^1, \ldots, S_F^p$) is AF-Stable and using implication, for all x^a, $a \in \{1, \ldots, p\}$, the following relation holds: $x^a(t) \in G^a(t)$ implies that $x^a(t + t_1) \in G^a(t + t_1)$ for all $t_1 > 0$. If the capture tubes are not safe, this implies that there exists $t_2 > 0$ such that x^a belongs to both tubes G^a and G^b, i.e., $x^a(t + t_2) \in G^a(t + t_2)$ and $x^a(t + t_2) \in G^b(t + t_2)$, $b \in \{1, \ldots, p\} \Rightarrow g^b(x^a, t + t_2) < 0$. Equivalently, the system of constraints in Eq. 4.74 is consistent, which is in contradiction with the assumption of the theorem. □

As a consequence, using AF-Stability and antifragile tubes, it is possible to establish the safety of a set of capture tubes. To our knowledge, no such algorithm exists. This is a very powerful result which can be applied in robotics to prove the safety of a squad of uncertain robots, i.e., no collision among robots. The safety will be illustrated in Sect. 4.6.1 for a squad of three robotic sailboats moving in their environment.

4.6 Applications Examples

4.6.1 Proving Stability for Trajectory-Tracking for a Sailboat Robot

In Sect. 4.6.1, a controller for a time-invariant sailboat robot will be proposed (for known wind direction) and then, using the theorems in the previous subsection, it will be proven that the controlled sailboat is AF-Stable. When the wind is unknown, the stability of the sailboat robot amounts to studying the stability of a differential inclusion by applying Theorem 4.2. In Sect. 4.6.1, the sailboat robot will track a time dependent trajectory and, in this case, the Theorem 4.4 will be used to prove that a target tube is an antifragile tube for the uncertain time-dependent sailboat robot, i.e., it is AF-stable.

Proving AF-Stability for time-invariant sailboat robot

Consider a sailboat robot described by the following velocity model:

$$\begin{pmatrix} x_1 \\ x_2 \end{pmatrix} = \sqrt{1 + \cos(\psi - u)} \cdot \begin{pmatrix} \cos u \\ \sin u \end{pmatrix}. \quad (4.75)$$

The input corresponds to the heading, θ, of the sailboat. In this model, ψ, is the measurable wind direction and $v = \sqrt{1 + \cos(\psi - u)}$ corresponds to the boat's speed. Figure 4.23 illustrates the corresponding polar speed diagram. Note that for $u \simeq \psi \pm \pi$ (which means that the sailboat robot is against the wind), the speed vanishes.

The proposed control law is influenced by the line following controller described in [86] where the sailboat robot follows the closed hauled angle taken as $\zeta = \frac{\pi}{4}$.

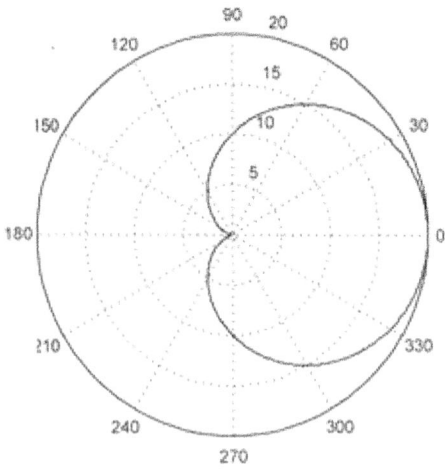

Fig. 4.23 Polar speed diagram of the sailboat robot with respect to $\psi - \theta$

4.6 Applications Examples

Denote the nominal angle by θ^*, which represents the desired direction. Due to the wind, θ^* may not be feasible (see Fig. 4.24). Hence, define $\bar{\theta}$ as the corrected angle. If θ^* is feasible, then $\bar{\theta} = \theta^*$, and when θ^* is not feasible, $\bar{\theta}$ is the nearest feasible angle. The proposed controller is given in Algorithm 8

Require: $X, \bar{X}, \psi, r$
Ensure: θ^*
  if $\|\bar{x} - x\| > \frac{r}{2}$ **then**
    $\theta^* = \text{atan} 2(\bar{x} - x)$
  end if
  if $\cos(\psi - \theta^*) + \cos \zeta < 0$ **then**
    $n = \pi + \psi + \text{sign}(\sin(\psi - \theta^*)) \zeta$
  else
    $n = \theta^*$
  end if return n

In the above algorithm, r is the radius of the bubble, and $\bar{x}$ (the desired position of the sailboat robot) is the center of the bubble. The bubble is the invariant set represented in Fig. 4.24. Studying the stability of the sailboat robot amounts to studying the stability of the time-invariant velocity model in Eq. 4.75, where u is given by Algorithm 8. The desired position is $\bar{x} = (20, 20)$, and the wind direction is represented by the thick black arrow in Fig. 4.24. Now, using Theorem 4.1, we will be able to prove AF-Stability of (4.75) for the radius of the bubble $r = 50$ m. The antifragile invariant set

$$V = \left\{ V(x) = (x_1 - \tilde{x}_1)^2 + (x_2 - \tilde{x}_2)^2 - r^2 < 0 \right\}$$

corresponds to the bubble with radius r and center $\tilde{x} = (\tilde{x}_1, \tilde{x}_2)$ (see Fig. 4.24). The properties of interest are the following:

Property 4.1 *If the sailboat robot has a distance to the desired position greater than $r = 50$ m then this distance will decrease until it reaches a value less than $r = 50$ m, i.e., $V(x) > 0$ then $\dot{V}(x) < 0$.*

Property 4.2 *If the robot has a distance to the desired position less than $r = 50$ m, then it will be the case forever. Inside the bubble of radius r, $\dot{V}(x) < 0$ for $\|\bar{x} - x\| > \frac{r}{2}$. For $\|\bar{x} - x\| < \frac{r}{2}$, the sailboat robot can have any trajectory, i.e., the sailboat robot will follow the wind direction since the controller is inactive. Figure 4.24 illustrates properties 1 and 2 for a particular wind direction represented with the thick black arrow.*

When the wind direction is unknown, i.e., ψ is unknown, then the controller proposed in Algorithm 8 becomes:

$$\begin{cases} u \in \text{atan} 2(\widehat{X} - X) + [-\zeta, \zeta] \text{ if } \|X - \widehat{X}\| > \frac{r}{2} \\ u \in [-\pi, \pi] \text{ otherwise} \end{cases}$$

Fig. 4.24 Controlled sailboat with controller from Algorithm 8 for one particular wind direction (the thick black arrow)

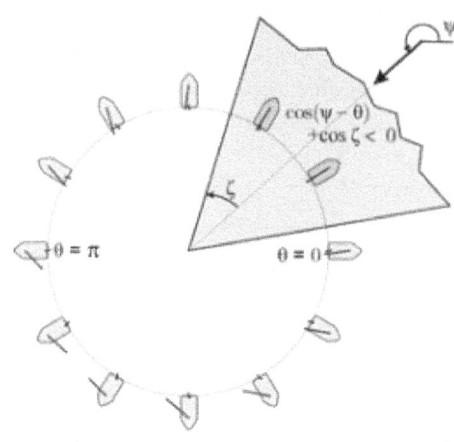

As a consequence, since the wind angle ψ is inside the interval $[\psi] = [-\pi, \pi]$, studying the stability of the sailboat robot amounts to studying the stability of the differential inclusion, i.e., an interval vector field

$$U : \begin{cases} R^2 \to IR^2 \\ (\cos(\arctan 2(\bar{x}) + [-\varsigma, \varsigma])) \\ \sin(\arctan 2(\bar{x}) + [-\varsigma, \varsigma])) \end{cases} \text{ if } \|\widehat{X}\| > \frac{r}{2}$$

where $\bar{x} = \widehat{x} - x$, $\bar{x}_1 = \widehat{x}_1 - x_1$, $\bar{x}_2 = \widehat{x}_2 - x_2$.

Applying Theorem 4.2 for the radius of the bubble $r = 50$ m one proves AF-Stability of the differential inclusion of Eq. 4.75. The antifragile invariant set $V = \{V(x) = (x_1 - \widehat{x}_1)^2 + (x_2 - \widehat{x}_2)^2 - r^2 < 0\}$ corresponds to the bubble with radius r and center $\widehat{x} = (\widehat{x}_1, \widehat{x}_2)$ (see Fig. 4.25). The properties of interest are the same as in the case when the wind direction is known:

Property 4.3 *If the sailboat robot has a distance to the desired position greater than $r = 50$ m then this distance will decrease until it reaches a distance less than $r = 50$ m for all possible wind directions, i.e., $V(x) > 0$ then $\dot{V}(x) < 0$.*

Property 4.4 *If the robot has a distance to the desired position less than $r = 50$ m, then it will be the case forever. Inside the bubble of radius r, $\dot{V}(x) < 0$ for $\|\widehat{x} - x\| > \frac{r}{2}$. For $\|\widehat{x} - x\| < \frac{r}{2}$ the sailboat robot can have any trajectory because $[\psi] = [-\pi, \pi]$. Figure 4.25 illustrates properties 1 and 2 when the wind direction is unknown. To draw this figure, we built a grid in the state space $x = (x_1, x_2)$ and for each x we have drawn arrows corresponding to all $\dot{x}$ consistent with the differential inclusion of Eq. 4.75, as it is illustrated in the magnified area in Fig. 4.25.*

Note that the C++ source code, the documentation, and examples for the solver Stabibex used to prove AF-Stability are available in [98].

4.6 Applications Examples

Fig. 4.25 Differential inclusion associated with the controlled sailboat using the controller for one wind

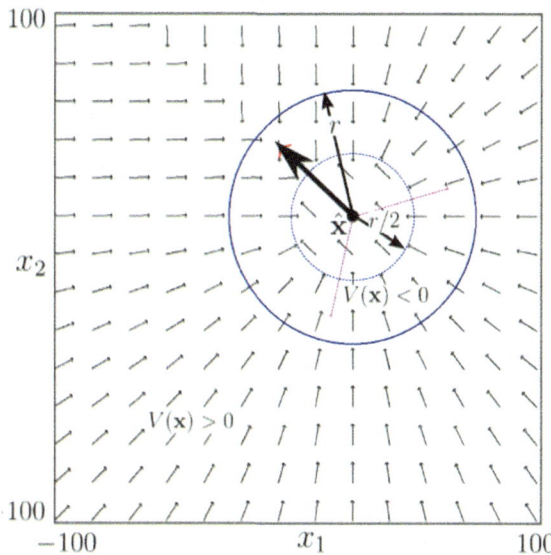

Proving AF-stability for time dependent sailboat robot

To prove the stability of the sailboat robot, we consider the target tube $G(t) = \{x, g(x,t) \leq 0\}$ for the differential inclusion of the model in Eq. (4.75), with $g(x,t) = (x_1 - \widehat{x}_1(t))^2 + (x_2 - \widehat{x}_2(t))^2 - r^2$, which corresponds to a time-dependent bubble with center $\widehat{x} = (\widehat{x}_1(t), \widehat{x}_2(t))$. The center is assumed to vary with time according to the following equation:

$$\begin{cases} \widehat{x}_1(t) = 175\cos(t) \\ \widehat{x}_2(t) = 87.5\sin(t) \end{cases}$$

which corresponds to an ellipsoidal trajectory. In this situation, the system becomes time-dependent, and the time-dependent bubble becomes an antifragile tube for the sailboat robot. Theorem 4.6 is used to prove that the time-dependent differential inclusion with the target tube $G(t) = \{x, g(x,t) \leq 0\}$ is AF-Stable. Therefore, it is necessary to prove that the target tube $G(t) = \{x, g(x,t) \leq 0\}$ is an antifragile tube for the sailboat robot. In consequence, the cross-out condition is verified for the tube $G(t) = \{x, g(x,t) \leq 0\}$ by applying Theorem 4.4. Since the tube is periodic with a period equal to 2π, the analysis can be restricted to $t \in [0, 2\pi]$. The cross-out condition is equivalent to show that the system of constraints

$$\begin{cases} (a)\ (x_1 - 175\cos(t))^2 + (x_2 - 87.5\sin(t))^2 - r^2 \leq 0 \\ (b)\ |u - \operatorname{atan2}(\tilde{x} - x)| \leq \zeta \\ (c)\ v = \sqrt{1 + \cos(\psi - u)} \\ (d)\ g_t = 350(x_1 - 175\cos(t))\sin(t) - 175(x_2 - 87.5\sin(t))\cos(t) \\ (e)\ g_x = 2v\cos(u)(x_1 - 175\cos(t)) + 2v\sin(u)(x_2 - 87.5\sin(t)) \\ (f)\ g_x + g_t \geq 0 \end{cases} \qquad (4.76)$$

is inconsistent for all $x = (x_1, x_2) \in R^2$, for all $\psi \in [-\pi, \pi]$, for all $u \in [-\pi, \pi]$, for all $t \in [0, 2\pi]$ and for $r = 50$ m. Equation (4.76)a corresponds to Conditions (iii) and (iv) of Theorem 4.4. Equation (4.76)c defines an intermediate variable v, which corresponds to the speed of the robot. Equation (4.76)d, e introduce two intermediate variables g_t and g_x which correspond to $\frac{\partial g}{\partial t}(x, t)$ and $\frac{\partial g}{\partial x}(x, t) f(x)$, respectively. Inequality (4.76)f corresponds to Condition (i) of Theorem 4.4. The system of constraints in Eq. (4.76) is inconsistent for $r = 50$ m then $G(t)$ is an antifragile tube and according with Theorem 4.6 the uncertain sailboat robot is AF-Stable. In this case, the property of interest is the following:

Property 4.5 *If all trajectories of the differential inclusion of Eq. (4.75) are inside the capture tube $G(t) = \{x, (x_1 - 175\cos(t))^2 + (x_2 - 87.5\sin(t))^2 - r^2 \leq 0\}$, it will be the case forever.* Figure 4.26 shows the AF-Stability for the sailboat robot for the ellipsoidal capture tube $G(t)$.

Proving the safety for a squad of three AF-Stable sailboat robots
Consider the problem of three sailboat robots R^1, R^2, R^3, each moving on a trajectory in their environment. The target trajectory has the form $\widehat{x}^i = \binom{h_1(t+\varphi_i)}{h_2(t+\varphi_i)}$, where $h = (h_1, h_2)$ is periodic of period T and $\varphi_i = \frac{T(i-1)}{3}$ is the time shift between the three targets. The target tube has the form $G(t) = \{x, g(x, t) \leq 0\}$ with $g(x, t) = \left(x_1^i - \widehat{x}_1^i(t)\right)^2 + \left(x_2^i - \widehat{x}_2^i(t)\right)^2 - r^2$ as in Sect. 4.6.1. In this case, if, for instance, the target trajectory is an ∞-shape trajectory, we have $\widehat{x}^i = \binom{k*\cos(t+\varphi_i)}{k*\sin(2(t+\varphi_i))}$, where $k > 0$. The sailboat robots are described by the differential inclusion of Eq. (4.75) since the wind direction is assumed to be unknown. It will be shown, using Theorem 4.7, that

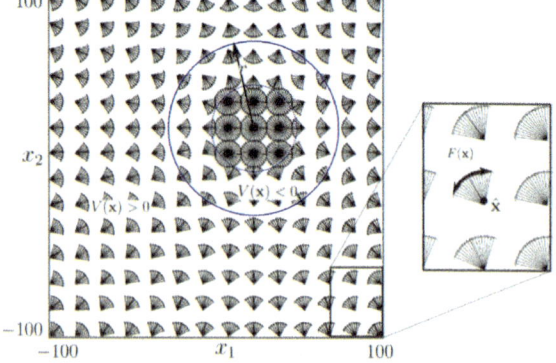

Fig. 4.26 AF-Stability for the sailboat robot for the ellipsoidal antifragile tube

4.6 Applications Examples

there is no collision among the robots in the squad, i.e., the squad is safe. It is, therefore, necessary to establish that for all $t > 0$ each robot is AF-Stable and the afferent antifragile tubes are safe (Definition 4.7). Moreover, for all $j \in \{1, 2, 3\}, i \neq j$, it will be proved, using Theorem 4.7, that for all $t > 0$, the set of inequalities

$$\begin{cases} (a) \ \left(x_1^i - \widehat{x}_1^i(t)\right)^2 + \left(x_2^i - \widehat{x}_2^i(t)\right)^2 - r^2 \leq 0 \\ (b) \ \left(x_1^i - \widehat{x}_1^j(t)\right)^2 + \left(x_2^i - \widehat{x}_2^j(t)\right)^2 - r^2 < 0 \end{cases} \quad (4.77)$$

is inconsistent.

Figures 4.27 and 4.28 show the squad of three sailboat robots following an ∞ shape antifragile tube with two different values of k. For this test-case $r = 50$, and the time shift between the three targets is $\varphi_1 = 0$, $\varphi_2 = \frac{3\pi}{3}$, and $\varphi_3 = \frac{4\pi}{3}$, respectively. In Fig. 4.27, for $k = 62.5$, the squad is not safe since the stability regions for the sailboat robots intersect for some time instants, i.e., the condition of Theorem 4.7 is not satisfied. In Fig. 4.28, for $k = 100$, the squad is safe since the stability regions for the sailboat robots never intersect, i.e., the system of constraints in Eq. (4.77) is inconsistent.

For clarity, the antifragile tubes are shown at a particular time instant in Figs. 4.27 and 4.28 and hence are represented by capture bubbles for the three sailboats. Note that the C++ source code, the documentation, and examples for the solver Bubbibex used to prove AF-Stability and safety are available in [90].

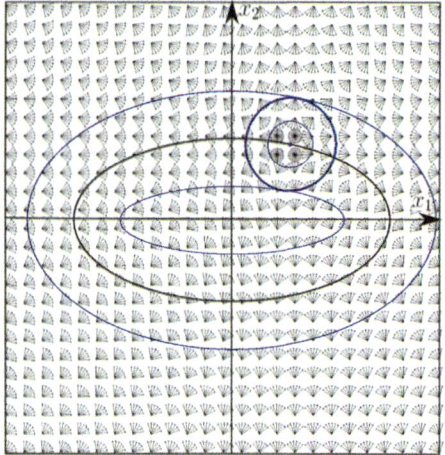

Fig. 4.27 The three capture tubes intersect for $k = 62.5$

Fig. 4.28 The three capture tubes never intersect for $k = 100$

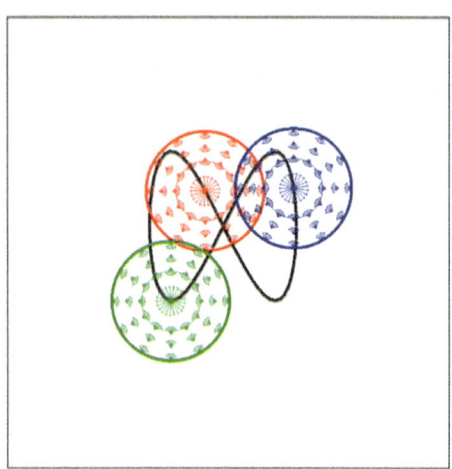

4.7 Conclusions

This chapter has considered the problem of determining the antifragile stability of nonlinear, time-dependent and uncertain systems in a set-membership context. It was proposed a simple and efficient method to solve the AF-Stability problem by proving the inconsistency of a set of inequalities. Moreover, the AF Stability was proved for time-dependent systems or time-dependent differential inclusions. To prove that a time-dependent system (or differential inclusion) is AF-Stable, it is necessary to find an antifragile tube for the system (or for the differential inclusion). To prove that a target tube is an antifragile tube is transformed into proving the inconsistency of a set of inequalities which can be efficiently solved, in polynomial-time, using a contractor-based method. To illustrate the concept of AF-Stability, a trajectory tracking application of an uncertain robotic sailboat was presented.

An important and yet unsolved problem for a squad of autonomous uncertain robots is to be able to prove that no collision occurs between the robots within the squad. To solve this problem, the concept of safety for a set of capture tubes has been proposed for the first time. This concept of safety is based on AF Stability and is established by proving the inconsistency of a set of inequalities. It is then used to analyze the safety of a squad of uncertain robotic sailboats (no collision among the robots).

References

1. van Wageningen-Kessels, F., Van Lint, H., Vuik, K., & Hoogendoorn, S. (2015). Genealogy of traffic flow models. *EURO Journal on Transportation and Logistics, 4*, 445–473
2. Hunt, P., Robertson, D., Bretherton, R., & Royle, M. C. (1982). The scoot on-line traffic signal optimisation technique. *Traffic Engineering & Control, 23*
3. Lowrie, P. (1990). *Scats, sydney co-ordinated adaptive traffic system: A traffic responsive method of controlling urban traffic.* Roads and Traffic Authority NSW, Traffic Control Section
4. Henry, J.-J., Farges, J. L., & Tuffal, J. (1984). The prodyn real time traffic algorithm. In *Control in Transportation Systems* (pp. 305–310). Elsevier
5. Gartner, N. H. (1983). *OPAC: A demand-responsive strategy for traffic signal control.* 906
6. Mirchandani, P., & Head, L. (1998). Rhodes-a real-time traffic signal control system: architecture, algorithm. In *TRISTAN III (Triennial Symposium on Transportation Analysis)*. volume 2
7. GmbH, S. . W. (2021). LISA+ traffic-planning software lisa
8. Chen, D. (2010). *Modeling travel time uncertainty in traffic networks*. Ph.D. thesis Massachusetts Institute of Technology
9. Ossenbruggen, P. J., Laflamme, E. M., & Linder, E. (2012). Congestion probability and traffic volatility. *Transportation research record, 2315*, 54–65
10. Ko, J., Guensler, R., & Hunter, M. (2006). Variability in traffic flow quality experienced by drivers: evidence from instrumented vehicles. *Transportation research record, 1988*, 1–9
11. Chedjou, J. C., & Kyamakya, K. (2018). A review of traffic light control systems and introduction of a control concept based on coupled nonlinear oscillators. *Recent Advances in Nonlinear Dynamics and Synchronization*, (pp. 113–149)
12. Axenie, C., Shi, R., Foroni, D., Wieder, A., Hassan, M. A. H., Sottovia, P., Grossi, M., Bortoli, S., & Brasche, G. (2021). Obelisc: Oscillator-based modelling and control using efficient neural learning for intelligent road traffic signal calculation. In *Joint European Conference on Machine Learning and Knowledge Discovery in Databases* (pp. 437–452). Springer
13. Li, J., & Zhang, H. M. (2011). Fundamental diagram of traffic flow: new identification scheme and further evidence from empirical data. *Transportation research record, 2260*, 50–59
14. Taleb, N. N., & Douady, R. (2013). Mathematical definition, mapping, and detection of (anti) fragility. *Quantitative Finance, 13*, 1677–1689
15. Axenie, C., Kurz, D., & Saveriano, M. (2022). Antifragile control systems: The case of an anti-symmetric network model of the tumor-immune-drug interactions. *Symmetry, 14*, 2034
16. Lee, J. M. (2006). *Riemannian manifolds: an introduction to curvature* volume 176. Springer Science & Business Media
17. Utkin, V. I. (2013). *Sliding modes in control and optimization*. Springer Science & Business Media
18. Slotine, J.-J. E., Li, W. et al. (1991). *Applied nonlinear control* volume 199. Prentice hall Englewood Cliffs, NJ
19. Levant, A. (1993). Sliding order and sliding accuracy in sliding mode control. *International journal of control, 58*, 1247–1263
20. Levant, A. (2005). Homogeneity approach to high-order sliding mode design. *Automatica, 41*, 823–830
21. Bartolini, G., Pisano, A., Punta, E., & Usai, E. (2003). A survey of applications of second-order sliding mode control to mechanical systems. *International Journal of control, 76*, 875–892
22. Blagodatskikh, V. I., & Filippov, A. F. (1985). Differential inclusions and optimal control. *Trudy Matematicheskogo Instituta Imeni VA Steklova, 169*, 194–252
23. Filippov, A. F. (1963). Differential equations with multi-valued discontinuous right-hand side. In *Doklady Akademii Nauk* (pp. 65–68). Russian Academy of Sciences volume 151
24. Song, J., Zuo, Z., & Basin, M. (2020). New class k_infinity function-based adaptive sliding mode control design. *arXiv preprint* arXiv:2012.02633
25. Kochetkov, S., Krasnova, S. A., & Utkin, V. A. (2022). The new second-order sliding mode control algorithm. *Mathematics, 10*, 2214

26. Owen, J., & Zames, G. (2020). Unstructured uncertainty in hinf. In *Control of Uncertain Dynamic Systems* (pp. 3–20). CRC Press
27. Jayasuriya, S. (2020). Frequency domain design for maximizing tolerance to disturbances in uncertain systems. In *Control of Uncertain Dynamic Systems* (pp. 215–231). CRC Press
28. Lopez, P. A., Behrisch, M., Bieker-Walz, L., Erdmann, J., Flötteröd, Y.-P., Hilbrich, R., Lücken, L., Rummel, J., Wagner, P., & Wießner, E. (2018). Microscopic traffic simulation using sumo. In *The 21st IEEE International Conference on Intelligent Transportation Systems*. IEEE
29. Gentile, Guido, Meschini, Lorenzo, & Papola, Natale (2007). Spillback congestion in dynamic traffic assignment: a macroscopic flow model with time-varying bottlenecks. *Transportation Research Part B: Methodological, 4*, 1114–1138
30. Ouyang, Y., Zhang, R. Y., Lavaei, J., & Varaiya, P. (2020). Large-scale traffic signal offset optimization. *IEEE Transactions on Control of Network Systems, 7*, 1176–1187
31. Strogatz, S. H. (2000). From kuramoto to crawford: exploring the onset of synchronization in populations of coupled oscillators. *Physica D: Nonlinear Phenomena, 143*, 1–20
32. Halder, P., & Althoff, M. (2022). Minimum-violation velocity planning with temporal logic constraints. In *IEEE 25th International Conference on Intelligent Transportation Systems (ITSC)*. https://ieeexplore.ieee.org/document/9922114
33. Haddadin S., R. A. u. T. S., Knobbe D. (2020). Geriatronik – assistenzroboter für ein selbstbestimmtes leben im alter? In M. Mokry, S. und Rückert (Ed.), *Roboter als (Er-)Lösung? Orientierung der Pflege von morgen am christlichen Menschenbild* Forschung - Technik – Praxis. Hans-Seidel-Stiftung
34. Chong, L., & Osorio, C. (2018). A simulation-based optimization algorithm for dynamic large-scale urban transportation problems. *Transportation Science, 52*, 637–656
35. Luca, A. D., & Oriolo, G. (1995). Modelling and control of nonholonomic mechanical systems. In *Kinematics and dynamics of multi-body systems* (pp. 277–342). Springer
36. Oriolo, G., De Luca, A., & Vendittelli, M. (2002). Wmr control via dynamic feedback linearization: design, implementation, and experimental validation. *IEEE Transactions on control systems technology, 10*, 835–852
37. Zhang, Y., Chung, J. H., & Velinsky, S. A. (2003). Variable structure control of a differentially steered wheeled mobile robot. *Journal of intelligent and Robotic Systems, 36*, 301–314
38. Solea, R., & Nunes, U. (2007). Trajectory planning and sliding-mode control based trajectory-tracking for cybercars. *Integrated Computer-Aided Engineering, 14*, 33–47
39. Axenie, C., & Solea, R. (2010). Real time control design for mobile robot fault tolerant control. introducing the artemic powered mobile robot. In *Proceedings of 2010 IEEE/ASME International Conference on Mechatronic and Embedded Systems and Applications* (pp. 7–13). IEEE
40. Kanayama, Y., Kimura, Y., Miyazaki, F., & Noguchi, T. (1990). A stable tracking control method for an autonomous mobile robot. In *Proceedings., IEEE International Conference on Robotics and Automation* (pp. 384–389). IEEE
41. Jiang, Z.-P., & Nijmeijer, H. (1997). Tracking control of mobile robots: A case study in backstepping. *Automatica, 33*, 1393–1399
42. Sarkar, N., Yun, X., & Kumar, V. (1994). Control of mechanical systems with rolling constraints: Application to dynamic control of mobile robots. *The International Journal of Robotics Research, 13*, 55–69
43. Samson, C., & Ait-Abderrahim, K. (1991). Feedback control of a nonholonomic wheeled cart in cartesian space. In *Proceedings. 1991 IEEE International Conference on Robotics and Automation* (pp. 1136–1137). IEEE Computer Society
44. Zhang, M., & Hirschorn, R. (1997). Discontinuous feedback stabilization of nonholonomic wheeled mobile robots. *Dynamics and Control, 7*, 155–169
45. Koh, K. C., & Cho, H. S. (1999). A smooth path tracking algorithm for wheeled mobile robots with dynamic constraints. *Journal of Intelligent and Robotic Systems, 24*, 367–385
46. Solea, R., Filipescu, A., & Nunes, U. (2009). Sliding-mode control for trajectory-tracking of a wheeled mobile robot in presence of uncertainties. In *2009 7th Asian Control Conference* (pp. 1701–1706)

References

47. Yang, J.-M., & Kim, J.-H. (1999). Sliding mode control for trajectory tracking of nonholonomic wheeled mobile robots. *IEEE Transactions on robotics and automation, 15,* 578–587
48. Kim, M.-S., Shin, J.-H., & Lee, J.-J. (2000). Design of a robust adaptive controller for a mobile robot. In *Proceedings. 2000 IEEE/RSJ International Conference on Intelligent Robots and Systems (IROS 2000)(Cat. No. 00CH37113)* (pp. 1816–1821). IEEE volume 3
49. Fukao, T., Nakagawa, H., & Adachi, N. (2000). Adaptive tracking control of a nonholonomic mobile robot. *IEEE transactions on Robotics and Automation, 16,* 609–615
50. Li, C., & Chao, H. (2002). Output tracking of uncertain robot systems via high order sliding mode technique. *Electronics Letters, 38,* 1
51. Wu, S.-F., Mei, J.-S., & Niu, P.-Y. (2001). Path guidance and control of a guided wheeled mobile robot. *Control Engineering Practice, 9,* 97–105
52. Jiang, Z.-P., Lefeber, E., & Nijmeijer, H. (2001). Saturated stabilization and tracking of a nonholonomic mobile robot. *Systems & Control Letters, 42,* 327–332
53. Pourboghrat, F., & Karlsson, M. P. (2002). Adaptive control of dynamic mobile robots with nonholonomic constraints. *Computers & Electrical Engineering, 28,* 241–253
54. Dong, W., & Kuhnert, K.-D. (2005). Robust adaptive control of nonholonomic mobile robot with parameter and nonparameter uncertainties. *IEEE Transactions on Robotics, 21,* 261–266
55. Axenie, C., & Cernega, D. (2010). Adaptive sliding mode controller design for mobile robot fault tolerant control. introducing artemic. In *19th International Workshop on Robotics in Alpe-Adria-Danube Region (RAAD 2010)* (pp. 253–259). IEEE
56. Saveriano, M., Franzel, F., & Lee, D. (2019). Merging position and orientation motion primitives. In *2019 International Conference on Robotics and Automation (ICRA)* (pp. 7041–7047)
57. Saveriano, M., & Lee, D. (2019). Learning barrier functions for constrained motion planning with dynamical systems. In *2019 IEEE/RSJ International Conference on Intelligent Robots and Systems (IROS)* (pp. 112–119)
58. Axenie, C., & Grossi, M. (2022). Antifragile control systems: The case of an oscillator-based network model of urban road traffic dynamics. *arXiv preprint*arXiv:2210.10460
59. Solea, R. C. (2009). *Sliding mode control applied in trajectory-tracking of WMRs and autonomous vehicles*. Ph.D. thesis Department of Electrical and Computer Engineering, University of Coimbra, Portugal
60. Fierro, R., & Lewis, F. L. (1997). Control of a nonholomic mobile robot: Backstepping kinematics into dynamics. *Journal of robotic systems, 14,* 149–163
61. Bloch, A. M. (2015). An introduction to aspects of geometric control theory. In *Nonholonomic mechanics and control* (pp. 199–233). Springer
62. Campion, G., d'Andrea Novel, B., & Bastin, G. (1991). Modelling and state feedback control of nonholonomic mechanical systems. In *[1991] Proceedings of the 30th IEEE Conference on Decision and Control* (pp. 1184–1189). IEEE
63. Godhavn, J.-M., & Egeland, O. (1997). A lyapunov approach to exponential stabilization of nonholonomic systems in power form. *IEEE Transactions on Automatic Control, 42,* 1028–1032
64. Walsh, G., Tilbury, D., Sastry, S., Murray, R., & Laumond, J.-P. (1994). Stabilization of trajectories for systems with nonholonomic constraints. *IEEE Transactions on Automatic Control, 39,* 216–222
65. Wit, C. C. d., Khennouf, H., Samson, C., & Sordalen, O. J. (1993). Nonlinear control design for mobile robots. In *Recent trends in mobile robots* (pp. 121–156). World Scientific
66. Solea, R., & Nunes, U. (2006). Trajectory planning with velocity planner for fully-automated passenger vehicles. In *2006 IEEE Intelligent Transportation Systems Conference* (pp. 474–480)
67. Fenichel, N. (1979). Geometric singular perturbation theory for ordinary differential equations. *Journal of differential equations, 31,* 53–98
68. Jones, C. K. (1995). Geometric singular perturbation theory. *Dynamical systems,* (pp. 44–118)

69. Hunter, J. K. (2004). Asymptotic analysis and singular perturbation theory. *Department of Mathematics, University of California at Davis*, (pp. 1–3)
70. Kokotović, P., Khalil, H. K., & O'reilly, J. (1999). *Singular perturbation methods in control: analysis and design*. SIAM
71. DeCarlo, R. A., Zak, S. H., & Matthews, G. P. (1988). Variable structure control of nonlinear multivariable systems: a tutorial. *Proceedings of the IEEE*, *76*, 212–232
72. Utkin, V. (1977). Variable structure systems with sliding modes. *IEEE Transactions on Automatic control*, *22*, 212–222
73. Utkin, V. I. (2008). Sliding mode control: mathematical tools, design and applications. In *Nonlinear and optimal control theory* (pp. 289–347). Springer
74. Utkin, V., & Yang, K. (1978). Methods for constructing discontinuity planes in multidimensional variable structure systems. *Automation and Remote control*, *39*, 1466–1470
75. Gao, W., & Hung, J. C. (1993). Variable structure control of nonlinear systems: A new approach. *IEEE transactions on Industrial Electronics*, *40*, 45–55
76. Hung, J. Y., Gao, W., & Hung, J. C. (1993). Variable structure control: A survey. *IEEE transactions on industrial electronics*, *40*, 2–22
77. Wang, H., Liu, B., Ping, X., & An, Q. (2019). Path tracking control for autonomous vehicles based on an improved mpc. *IEEE Access*, *7*, 161064–161073
78. Antonelli, G., Chiaverini, S., & Fusco, G. (2007). A fuzzy-logic-based approach for mobile robot path tracking. *IEEE transactions on fuzzy systems*, *15*, 211–221
79. Lyapunov, A. The general problem of motion stability. *Annals Of Mathematics Studies*. **17** (1892)
80. Tucker, W. The Lorenz attractor exists. *Comptes Rendus De L'Académie Des Sciences-Series I-Mathematics*. **328**, 1197–1202 (1999)
81. Malti, R., Moreau, X., Khemane, F. & Oustaloup, A. Stability and resonance conditions of elementary fractional transfer functions. *Automatica*. **47**, 2462–2467 (2011)
82. Jaulin, L. & Le Bars, F. An interval approach for stability analysis: Application to sailboat robotics. *IEEE Transactions On Robotics*. **29**, 282–287 (2012)
83. Singh, A. & Khalil, H. Regulation of nonlinear systems using conditional integrators. *International Journal Of Robust And Nonlinear Control: IFAC-Affiliated Journal*. **15**, 339–362 (2005)
84. Milanese, M., Norton, J., Piet-Lahanier, H. & Walter, É. Bounding approaches to system identification. (Springer Science & Business Media,2013)
85. Lhommeau, M., Jaulin, L. & Hardouin, L. Capture basin approximation using interval analysis. *International Journal Of Adaptive Control And Signal Processing*. **25**, 264–272 (2011)
86. Kailath, T. Linear systems. (Prentice-Hall Englewood Cliffs, NJ,1980)
87. Aubin, J. & Frankowska, H. Set-valued analysis, viability theory and partial differential inclusions. (WP-92-060,1992
88. Moore, R., Kearfott, R. & Cloud, M. Introduction to interval analysis. (SIAM,2009)
89. Sauze, C. & Neal, M. An autonomous sailing robot for ocean observation. (2006)
90. Jaulin, L. Programs and demos for V-stability and G-stability (2014) www.ensta-bretagne.fr/jaulin/mer.html
91. Aubry, C., Desmare, R. & Jaulin, L. Loop detection of mobile robots using interval analysis. *Automatica*. **49**, 463–470 (2013)
92. Cruck, E., Moitie, R. & Seube, N. Estimation of basins of attraction for uncertain systems with affine and Lipschitz dynamics. *Dynamics And Control*. **11**, 211–227 (2001)
93. Saint-Pierre, P. Approximation of the viability kernel. *Applied Mathematics And Optimization*. **29** pp. 187–209 (1994)
94. Berge, C. Topological spaces: Including a treatment of multi-valued functions, vector spaces and convexity. (Oliver & Boyd,1877)
95. Kurzhanski, A. & Vályi, I. Ellipsoidal calculus for estimation and control. (Springer,1997)
96. Yu, W., Chen, G. & Cao, M. Some necessary and sufficient conditions for second-order consensus in multi-agent dynamical systems. *Automatica*. **46**, 1089–1095 (2010)

References

97. Le Bars, F., Sliwka, J., Jaulin, L. & Reynet, O. Set-membership state estimation with fleeting data. *Automatica.* **48**, 381–387 (2012)
98. Jaulin, L., Kieffer, M., Didrit, O., Walter, E., Jaulin, L., Kieffer, M., Didrit, O. & Walter, É. Interval analysis. (Springer,2001)
99. Agrawal, & et al (1998). Non-linear control strategies for Duffing systems. *International Journal of Non-Linear Mechanics, 33*, 829–841

Chapter 5
Conclusions and Open Research Questions

Abstract This chapter concludes our interdisciplinary endeavour into technical systems antifragility. Be it intrinsic, inherited or induced, antifragility expands and consolidates the typical analysis, modelling, and design of technical systems. We hereby offer the main concepts that define antifragility levels in technical systems as well as the new research avenues our insights offer.

5.1 Intrinsic (Anti)-Fragility

Intrinsic antifragile systems benefit from the in-homogeneity of the internal dynamics distribution based on the convexity of the response function of the system without external input and solely based on the heterogeneity and resilience of the internal components. Characteristics such as stability describe the simplest system response with minimal antifragile properties. Within this scale, the precise characterization of the payoff function, which describes the relationship between system inputs and outputs, is of paramount importance.

In considering intrinsic antifragility, it is essential to understand the dynamics of timescale separation, which describes the input-output coupling. An illustrative example is that of urban intersections with varying characteristics (road width), where there is a significant discrepancy in delays even in instances where the traffic conditions remain relatively stable, due to fluctuations in arrival and discharge rates.

The empirical evidence indicates that delays increase in line with traffic flow and are constrained by a maximum flow number. As the flow approaches the boundary, the delays increase exponentially with the flow, which demonstrates empirically that the system is fragile. For instance, the convex shape of the traffic control payoff functions shows that the minimization of delays is achieved through a decrease in variability rather than an increase. Management strategies [3] are designed to drive the system towards critical density. These strategies take into account the impact of input variability (traffic flow) on the outcome (delays), which is influenced by the shape of the input-output function (exponential) [4].

In another example, the interactions between multiple control loops (e.g., the internal DC motor control loop and outer robot position control loop, the oscillators models of road traffic flow inputs, and the coupled traffic dynamics) are quantified in the context of input-output mappings of timescale separation, which are performed to handle uncertainty and high-frequency phenomena [5].

It should be noted, however, that such systems employ intrinsic antifragility to achieve their prescribed objectives by design, given the specifications and constraints imposed by the physics of the operational space of the system [2]. This is exemplified by driver models or road traffic dynamics [6].

The term 'criticality' is used to describe the onset of unpredictable fluctuations in a dynamical system that is on the verge of losing its stability [7]. The capacity to absorb and respond to stresses resulting from the emergence of scale-free temporal fluctuations, slowing dynamics, and multistability serve as the fundamental indicators of criticality. In this context, antifragility can be conceptualized as the motion of a system from an existing steady state to a superior one in the aftermath of a change in conditions. This transition among steady states may be continuous (second-order phase transition) or discontinuous (first-order phase transition).

5.2 Inherited (Anti)-Fragility

There is a positive correlation between antifragility and increased system heterogeneity. This aspect is pertinent when considering the system's capacity to build additional capacity in anticipation of disturbances. To achieve inherited antifragility, a system designer must build upon the timescale separation of a redundant overcompensation component. This concept has been demonstrated in several applications, including robotics [5], traffic control [4], and medical procedures [8].

Similarly, machine learning systems maintain timescale separation through the formulation of the learning task. In more precise terms, the system determines the optimal action that will yield the greatest possible discounted future reward. The contribution of the timescale separation and the redundant overcompensation term was isolated, and it was observed that the machine learning inherited antifragility controller gained in the skewness (i.e., convexity) of the disruption magnitude over surging demand. In such systems, the degree of antifragility is quantified in terms of the geometric properties of the anticipated control actions, as reflected in the shape of the response to high-magnitude external perturbations.

In addition, criticality plays a role in inherited antifragility by enabling a system to transition from one steady state to another. The trigger for state switching may originate from a change in the (1) parameters of the system, (2) externally introduced noise, or (3) an increase in the rates of the system from neighboring entities in a competition of cooperation with the system.

5.3 Induced (Anti)-Fragility

The control-theoretic approach to induced antifragility is based on the combination of timescale separation and redundant overcompensation with variable structure control. A controller steers the system towards the antifragile region of its operational domain by selecting an external control or regulation signal strategically.

This can be achieved by developing a control law that incorporates a redundant overcompensation capacity to handle potential sensor and actuator failures. By pushing the closed-loop system dynamics to prescribed limits, the control law can ensure the system remains stable and reliable, even in uncertain environments. This is illustrated, for instance, by the fragile–antifragile behaviour of a robot in uncertain environments across spatiotemporal dynamics [5]. Antifragility is quantified in terms of the quality of the dynamics tracking and the speed of reaching the desired region of the desired dynamics manifold in the presence of uncertainty and volatility. This is achieved through the use of adaptive control, robust control and resilient control strategies. In a large-scale traffic control application, the antifragile controller demonstrated statistically significant gains in the face of increasing traffic disruptions over time. The systematic evaluation demonstrates that selecting a control law based on the second-order effects of signal re-computation can effectively capture the volatile dynamics of the closed-loop system [4].

Machine learning techniques can be employed to induce antifragility, as demonstrated by the design of a traffic reinforcement learning agent that learns to be conservative when regulating the controlled region. Once again, we have a clear quantification of the system's antifragility based on a dynamics response curve to external uncertainty (i.e. the amplitude of traffic disruptions) and volatility (i.e. the onset and offset of traffic disruptions). This approach overcomes the baseline approach (i.e. static police-made traffic light control), a state-of-the-art model predictive control, and other reinforcement learning approaches [3].

Finally, when considering criticality, the control of multistability is about managing the transition of the system to a more desired steady state and preventing it from moving to an inferior one. It is beneficial for a controller to anticipate tipping points well before they occur so that remedial actions can be taken promptly.

References

1. Axenie, C., López-Corona, O., Makridis, M., Akbarzadeh, M., Saveriano, M., Stancu, A. & West, J. Antifragility in complex dynamical systems. *Npj Complexity*. **1**, 12 (2024)
2. Axenie, C., Scherr, W., Wieder, A., Torres, A., Meng, Z., Du, X., Sottovia, P., Foroni, D., Grossi, M., Bortoli, S. & Others Fuzzy modelling and inference for physics-aware road vehicle driver behaviour model calibration. *Expert Systems With Applications*. **241** pp. 122590 (2024)
3. Sun, L., Makridis, M., Genser, A., Axenie, C., Grossi, M. & Kouvelas, A. Antifragile Perimeter Control: Anticipating and Gaining from Disruptions with Reinforcement Learning. (arXiv,2024,2), http://arxiv.org/abs/2402.12665, arXiv:2402.12665 [cs, eess]

4. Axenie, C. & Grossi, M. Antifragile Control Systems: The case of an oscillator-based network model of urban road traffic dynamics. (arXiv,2023,1), http://arxiv.org/abs/2210.10460, arXiv:2210.10460 [cs, eess]
5. Axenie, C. & Saveriano, M. Antifragile Control Systems: The case of mobile robot trajectory tracking in the presence of uncertainty. *IEEE Access*. **11** pp. 138188-138200 (2023), http://arxiv.org/abs/2302.05117, arXiv:2302.05117 [cs, eess]
6. Makridis, M., Mattas, K., Ciuffo, B. & Kouvelas, A. Impacts of Partially Connected and Automated Vehicles on Traffic Flow and Energy Based on Worldwide Experimental Observations in Motorway Driving. *Sustainable Automated And Connected Transport*. **19** pp. 23–45 (2024,1), https://doi.org/10.1108/S2044-994120240000019002
7. Akbarzadeh, M., Memarmontazerin, S., Derrible, S. & Salehi Reihani, S. The role of travel demand and network centrality on the connectivity and resilience of an urban street system. *Transportation*. **46** pp. 1127–1141 (2019)
8. Axenie, C., Kurz, D. & Saveriano, M. Antifragile Control Systems: The Case of an Anti-Symmetric Network Model of the Tumor-Immune-Drug Interactions. *Symmetry*. **14**, 2034 (2022,10), https://www.mdpi.com/2073-8994/14/10/2034, Number: 10 Publisher: Multidisciplinary Digital Publishing Institute

The manufacturer's authorised representative in the EU is Springer Nature Customer Service Centre GmbH, Europaplatz 3, 69115 Heidelberg, Germany. If you have any concerns regarding our products, please contact ProductSafety@springernature.com

Printed and bound by CPI Group (UK) Ltd, Croydon, CR0 4YY

26/03/2026

02078954-0002